The Sentient Cell

Advanced Praise for *The Sentient Cell: The Cellular Foundations of Consciousness*

The perfect book for any scientifically curious individual seeking updates to research on my favourite subject in all the world—the intelligent living cell.

Brian J. Ford
President Emeritus
University of Cambridge Society for the Application of Research (UK)

The most comprehensive vision of post-neo-Darwinian biology to date. Life is not about genes and digital information codes. Life is about sentience, consciousness, mind, and intelligence. The building blocks of sentience and consciousness are cells. A landmark book that heralds a welcome paradigm shift in biology.

Predrag B. Slijepčević
Department of Life Sciences, Brunel University London (UK)

In 20 years of researching biological cognition in unicellular organisms I have studiously avoided what I call the 'C-word', consciousness. *The Sentient Cell* has made me thoroughly reconsider my reticence. It is a remarkable book that exposes the standard model of consciousness for the early 19th-century relic it is, providing abundant empirical evidence of why it should be supplanted. An excellent introduction to a necessary conversation that will unfold over the coming decades and which sits squarely on the right side of history. Vive la révolution!

Pamela Lyon
University of Adelaide (Australia)

There are few questions as fascinating, far-reaching, and impactful as the emergence of multiscale minds from the competencies of physico-chemical systems. This highly thought-provoking book addresses the continuity of life and diverse intelligence across evolution, physiology, and behaviour. An engaging tour of ideas essential for understanding our nature and our biological, artificial, and hybrid futures.

Michael Levin
Vannevar Bush Professor
Biology Department, Tufts University (USA)

Tackles head-on and without apology the ultimate question: what are the very foundations of sentience and how did it spread throughout the tree of life? Leave aside any deep-rooted, human-centred prejudices, and fasten your seat belt. Be humble, the woven expertise on display is well worth the ride!

Paco Calvo
Minimal Intelligence Lab, University of Murcia (Spain)

This new book presents a sentient, dynamic, eternal boundary, dividing and defining the inner and outer world of cells, translating and interpreting between both, steering their evolution and supplying meaning to life.

Anton Markos
Department of Philosophy and History of Science,
Charles University (Czech Republic)

Although the view that all organisms possess consciousness is widespread, it is rarely discussed, let alone defended by Western academics. *The Sentient Cell* champions the position that all life is sentient, exploring even its most controversial consequences, including plant consciousness and its ethical implications. The book is consistently provocative, erudite, and engaging

Robert N. McCauley
William Rand Kenan Jr. University Professor Emeritus
Center for Mind, Brain, and Culture, Emory University (USA)

In *The Sentient Cell*, readers are granted an extraordinary view into the sophisticated nature of cells and the complex tapestry of life. Remarkably, numerous technologies once thought to be recent human inventions have been in existence for over a billion years, thanks to the clever biomolecular craft of microorganisms. This book promises to transform the way you perceive the world.

Perry Marshall
Author, *Evolution 2.0*

With cognition evident at all scales in nature, animals large and small, plants, and unicellular creatures, a new synthesis of many different scientific fields is necessary to understand how this can occur and what it might mean in relation to our human consciousness. This collaboration of three eminent scientists from very different vantage points makes a remarkable leap in our understanding of this complex project.

Jon Lieff, MD
Neuropsychiatrist
Author, *The Secret Language of Cells*

The Sentient Cell

The Cellular Foundations of Consciousness

Arthur S. Reber, František Baluška,
William B. Miller Jr.

OXFORD
UNIVERSITY PRESS

OXFORD
UNIVERSITY PRESS

Great Clarendon Street, Oxford, OX2 6DP,
United Kingdom

Oxford University Press is a department of the University of Oxford.
It furthers the University's objective of excellence in research, scholarship,
and education by publishing worldwide. Oxford is a registered trade mark of
Oxford University Press in the UK and in certain other countries

First Edition published in 2023

Published in the United States of America by Oxford University Press
198 Madison Avenue, New York, NY 10016, United States of America

British Library Cataloguing in Publication Data

Data available

Library of Congress Control Number: 2023936360

ISBN 978–0–19–887321–1

DOI: 10.1093/oso/9780198873211.001.0001

Preface

This volume is the natural follow-up to Arthur Reber's 2019 book, *The First Minds: Caterpillars, 'Karyotes, and Consciousness* (TFM). In that earlier work, the Cellular Basis of Consciousness[1,2] (CBC) theory was developed based on a number of earlier efforts published in a variety of journals between 1997 and 2019 as well as in talks, colloquia, and presentations at conferences. The core proposition in TFM was that life and mind are coterminous. All organisms, all species extant and extinct, are sentient. All have an existentially secure consciousness—without which they would have been evolutionary dead ends, unable to survive in the chaotic, dangerous environment in which life first appeared. And, importantly, all forms of sentience, all forms of cognitive functioning right up to and including those expressed by humans, evolved from the original expression of consciousness at the birth of life in prokaryotes. The proposition that all life forms evolved from those first unicellular species is a widely accepted, foundational principle of the biological and social sciences. The CBC model simply applies that same proposition to sentience.

What was missing from those efforts was an in-depth exploration of the underlying biochemical, biomolecular, and microbiological factors that were/are responsible for creating sentient cells. The reason was simple. Reber lacked the background, training, and experience in the biological sciences needed for such an exploration. The obvious route was to reach out to others with the requisite knowledge and skills. František Baluška had already been on board in the sense that he was one of the cell biologists to whom Reber reached out for advice while working on TFM. They had also gotten to know each other as participants in the Summer Institute on the 'Other Minds' problem held at the Université du Québec à Montréal in June 2018. It was at this conference

[1] A note of appreciation and apology to Lynn Margulis and Pamela Lyon whose important papers (Margulis, 2001 and Lyon, 2015) were respectively titled 'The conscious cell' and 'The cognitive cell'. We borrowed their approach here but note that our title has the advantage of being alliterative.

[2] Before we begin our deep dive into the origins of minds, a word on terminology. *Consciousness*, in the words of cognitive scientist George Miller, is 'a word worn smooth by a million tongues'. Synonyms such as 'sentience', 'cognition', 'mentation', and a host of others that can be found dotted throughout the literature have also been similarly worn down to lexicographic nubs. We will not try to define them but, rather, use them in a folk psychology fashion, as a collation of rough synonyms—rather like they are handled by the lay public. See Appendix I for an historical overview of relevant terminology and why we chose this lexicographic route and Appendix II for a glossary of technical terms from micro-biology, plant neurobiology, and the cognitive neurosciences that are sprinkled throughout the text.

that the decision to work together on this book was reached. Their approaches were so similar that Reber recalls thinking 'Oh no, he's giving *my* talk' while Baluška was presenting. Bill Miller, a productive medical researcher and frequent collaborator of Baluška's, was the obvious choice to round out the team. We put together a prospectus and sent it to Oxford University Press, the publisher of TFM and Reber's 1993 book. After extensive external reviews, they agreed to publish.

One thing was clear from the beginning. Because we were a bit of a patchwork team coming from very different scientific backgrounds, we needed to distribute responsibilities for material that reflected our areas of expertise. Reber's degree is in experimental psychology and his primary research focus has been on cognitive processes, in particular 'implicit learning'—where knowledge and skills are learned without explicit knowledge of either the processes or products of acquisition. Think of how infants learn language and how we all become socialized and come to learn how to behave appropriately in a culture to get a feeling for the areas where these processes function. Recently, Reber and Rhianon Allen edited and wrote parts of *The Cognitive Unconscious: The First Half-Century* (2022) in which over 30 scholars review the research in a wide array of topics that emerged from the early experiments on implicit cognition.

Baluška's degree is in plant physiology and plant cell biology, focusing on polarity, cytoskeleton, endocytosis, and vesicle recycling as well as other basic aspects of biology of cells. He is one of the founders of the sub-field of plant neurobiology where his insights into the causal role that cognition plays in the life of plants has become increasingly influential in recent years.

Miller's original training is in medicine and his interests have long been in developing novel theories for cellular function that expanded the causal factors that operate in evolution. He has consistently argued that most contemporary models of evolution are seriously lacking in explanatory power, primarily because they focus too tightly on genetic factors and fail to take into account a host of other variables such as information, choice, epigenetics, the impact of an error-prone biology, the role of stochastic processes, and, of course, cognition.

What emerged from our collaboration is what we regard as a revolutionary framework for viewing not just consciousness but a model that offers a genuinely new perspective on evolution and the nature of life on this planet. We provide support for a framework that downplays the role of genes by making clear that epigenetic and senomic elements play critical roles, emphasizes the importance of information and how it is processed by all living forms, embraces continuity across life forms, integrates all species into a singular framework, and provides a theoretical platform to understand how cognitive functions evolved after putting in its original appearance some 4 billion years ago. Keep in mind that all life was unicellular for some 2 billion of those years. The implication of this simple fact is that the terms 'cell' and 'organism' denoted identical entities, a point that is often missed when the issues raised by our CBC theory are contemplated. The organismal, sentient nature of cells, all cells including those in our human bodies, is still valid. We also recognize that our approach raises a number of questions that touch on ethical conduct and moral behaviour. These are discussed at length in Chapter 12.

Unlike virtually all other approaches to the broad topic of consciousness, we begin with the simplest species, with the first appearance of life, and examine the development and progression of cognitive functions across the evolutionary tree. We go into some detail outlining why the decision on the part of other scientists and philosophers to begin the explorations with *Homo sapiens* was a tactical error.

Will we convince others? We do not know, of course, but we hope that, as the advantages of taking the CBC model as the framework for an encompassing narrative of evolutionary biology become understood, we will see a paradigm shift. Amusingly, all of us have had the opportunity to present our views to various groups. A pattern has emerged—a predictable one. Groups of intelligent, curious laypersons generally respond to our insistence that life begins with intelligence with comments like, 'Now that's interesting. I hadn't thought of it, but it does make sense.' Cognitive scientists and neuroscientists tend to listen but question and most conclude that our position is a bridge too far. Philosophers, in particular philosophers of mind, typically think we have lost ours. But some cell biologists, ones who study bacteria and botanists with close ties to the plant neurobiology sub-field who examine the cognitive nature of plant life, respond with 'Well, duh. Of course.'

We rather like the recent move in many scientific journals to outline who was responsible for what. Here is ours. Reber undertook writing the initial drafts of the Prologue and Chapters 1, 2, 12, and this Preface. Parts of Appendix I were imported from TFM (with modifications and, of course, permission) and updated by Reber. Baluška wrote the first drafts of Chapters 4, 5, 6, 10, and 11 and coordinated with Miller in compiling the glossary in Appendix II. Miller took on responsibility for Chapters 3, 7, 8, and 9. Everything, of course, was read, edited by others, sent back, rewritten, edited again, and fine-tuned based on the feedback. It all went remarkably well with virtually never a harsh word or serious disagreement about how a topic or issue was being handled.

Even more remarkable, the volume was completed within the initial time frame. It is possible to pull this off when you have such respect for each other.

We do need to warn readers that some of the material, especially in the chapters that focus on the underlying biomolecular processes that 'create' sentience, can be a bit technical. Consult Appendix II when needed. We tried to tone down the presentation where we could but, of course, did not want to do violence to the science. The natural processes that occurred when 'chemistry made biology' as Addy Pross, an evolutionary biochemist, put it, was certainly not easy and the even more complex one of prokaryotes making eukaryotes was even more so—as the timeline reveals. It would be unwise to assume that it all could be expressed without the technical details.

Our backgrounds and current scientific efforts can be found in our web pages and other Internet locations.

For Reber:
https://psych.ubc.ca/profile/arthur-reber/
https://en.wikipedia.org/wiki/Arthur_S._Reber

For Baluška:
https://www.thethirdwayofevolution.com/people/view/frantisek-baluska
http://ds9.botanik.uni-bonn.de/zellbio/AG-Baluska-Volkmann/

For Miller:
https://humansandnature.org/william-b-miller-jr/
https://williammillermd.com

Prologue: Setting the Stage

Introduction

As the title of our compact volume states, our goal is to make sense of the classic problem of the emergence of sentience, consciousness. A few years ago, the American Association for the Advancement of Science (AAAS) considered this to be the second most important unanswered question in science. 'What is the universe made of?' was rated as the most important. Before getting into the theoretical and empirical details of our approach, we need to make our central proposition clear. Our view is that sentience, consciousness, self-referential awareness—whatever term(s) are used, and we will have more to say on terminology below—is an inherent feature of all living organisms. In over a baker's dozen chapters and appendices, we will advance arguments based on foundational principles of evolutionary biology and provide empirical evidence from recent research in microbiology to support the proposition. Put simply, life and sentience are coterminous.

As noted in the Preface, several chapters present the underlying biomolecular functions of cells. This material is designed to explain the basic cell biology processes that support cellular sentience and, necessarily, is technical. Appendix II provides definitions of the technical terms from microbiology. Consult freely. Philosopher David Chalmers famously dubbed the question of 'How do brains make minds?' as the Hard Problem of consciousness. What these chapters do is provide the cellular biomolecular answer to the rephrased problem, 'How do cells make sentience?'

Lexicography, Language, and Labels

Before embarking on our voyage, we need to confront some lexicographic issues. There is a good deal of confusion about what core terms like 'consciousness' or 'sentience' mean or what kinds of cognitive and affective functions are designated by them. To prevent misunderstanding, we will note here, at the beginning of our exegesis, that each of these terms will be used interchangeably and flexibly to emphasize one or another attribute of consciousness. We will be using these terms, and a variety of synonyms and near synonyms, in what's generally known as a 'folk psychology' manner—how the typical intelligent layperson treats them. In Appendix I (which was imported, with modifications (and permission), from Reber's 2019 book *The First Minds*), we cover the history and manner of use in a variety of overlapping fields and point out the advantages of the terminological position we've taken. We want to prevent what we see happening in the literature on a regular basis. A paper, authored by respected scholars, presents behavioural or biomolecular evidence that supports the notion that a species once assumed to be insentient does, in fact, possess genuine consciousness. Another paper, authored by equally respected researchers, criticizes it claiming that what the original authors identified is not 'really consciousness'. In the next chapter we review several instances where this is precisely what happened. We also anticipate that reviewers of this book will do so as well. It becomes a lexicographic debate, not a scientific one. Nobody wins and progress is thwarted.

We regard consciousness as a cohesive suite of cognitive and affective faculties encompassing valenced experiences, self-referencing, self-awareness, organized information assessment, communication, learning, memory, decision-making, and preference formation all linked to problem-solving in real-world settings. When viewed within these defining particulars, all life is sentient life for all species express these functions to some extent. As we will outline in Chapter 2, even unicellular species have valenced experiences, learn, form surprisingly stable memories, communicate meaningfully with each other, and are capable of contingent decision-making and problem-solving. At all scales, every form of life is not merely conscious in some vague sense but is specifically capable of self-referential experiences.

We recognize that our view of life and sentience being coterminous is not the one generally accepted in the larger field of what philosopher Ned Block calls 'consciousness science'. The standard paradigm is to treat consciousness as a feature of relatively sophisticated species, in particular, ones with a nervous system, and to view human mental life as the ultimate instantiation of it on the planet. There are reasons why the current paradigm is the dominant one. It has significant historical, epistemic, and cultural roots and we will discuss them in later chapters while pointing out the advantages of taking our cell-based position.

Different Platforms for Sentience

The argument that all but only living organisms are sentient and exclusively so, has, of course, an important implication: artificial entities, computers, robots, and digital computational devices are not conscious in the biological sense and, almost certainly never will be. We appreciate that it may not be wise to proclaim that something that intelligent people can imagine is possible, is *im*possible (that is why we put the 'almost' in there). We are not dogmatic—but we are empirical. If evidence for sentience expressed by an artificial intelligence (AI) based on digital operations is presented, we are more than willing to change our minds. We are appreciative of the fact that computing devices have been assembled by humans in just the last hundred years or so and AI, as a field of research, has only been around for a few decades. Living systems are the products of biological evolution which has been going on for some 4 billion years. There's time for the artificial ones to 'evolve'.

However, existing artificial systems can hoodwink the unwary. Recently, Blake Lemoine, an engineer at Google, was part of a team working on a project called the Language Models for Dialog Applications (LaMDA). The goal of the project is to design an AI that can engage in realistic conversations with humans on a wide variety of topics—unlike chatbots[3] which are restricted to specific issues. Ironically, Lemoine became a victim of a sophisticated bit of software that he helped develop. LaMDA is based on code that uses Google's 'Transformer' neural network and is designed to mimic the ways in which humans converse with each other. The Transformer platform is now open source and is widely used to create chatbots for many agencies and businesses. We have all been in conversations with digital chatbots—though we may not realize it. They can be quite effective.

After conversing extensively with LaMDA, Lemoine concluded that it was, in fact, sentient, had emotions, and could empathize with others—in short, that it had passed the Turing Test[4] and was conscious. He sent around an

[3] Chatbots are applications that conduct online conversations through text or text-to-speech. They are used widely to engage with customers and answer questions that they have about products or issues they may have with a company and, of course, are used by marketers on what are generally known as 'robocalls'. The open-source AI, ChatGPT, is uncanny in its ability to answer questions, generate novel recipes based on a simple request, write essays, and, in ways troublesome to teachers, write term papers. Like LaMDA, it does stumble over itself when the topic is one it has had little contact with and, of course, it doesn't really know anything. In Thomas Nagel's way of viewing consciousness (Nagel, 1974), there is nothing it is like to be ChatGPT.

[4] British mathematician and computer scientist Alan Turing (1912–1954) was one of the founders of the field of AI. He maintained that the true test of a sentient AI would be when a human was communicating with both another human in one room and a computer in another and could not determine which one was the other human. Many existing AIs can pass a limited version of the test, one where the domain of questioning is restricted to specific topics. When there are no limits placed on the topics, the artificial entity is easily identified.

'interview' he and a colleague had with it as evidence to support his contention. Lemoine introduces himself on his blog with: 'I'm a software engineer. I'm a priest. I'm a father. I'm a veteran. I'm an ex-convict. I'm an AI researcher. I'm a cajun [*sic*]. I'm whatever I need to be next.' But he neglected to add 'I'm credulous.' His blog does have thousands of followers and his post of the interview went viral. Predictably, it was picked up by major media outlets which, in several cases, presented his claims uncritically. Cable shows, including the popular *The Daily Show*, then with Trevor Noah as host, had a segment on it—though the credulity was balanced with comedy.

Google, however, was not amused. Their view of LaMDA is that it is a sophisticated digital device, an AI, one that does mimic humans in conversations—provided the necessary information has been fed in. Consciousness is not considered to be one of its features. LaMDA was trained on real-world dialogue and a huge database was created over many years by entering snippets of conversations. The result was an AI that 'produces a model that can be trained to read many words (a sentence or paragraph, for example), pay attention to how those words relate to one another and then predict what words it thinks will come next'.[5] You can, as Lemoine did, ask it what it thought of *Les Misérables*. LaMDA responded appropriately and they had an engaging discussion of the book and the Broadway musical based on it. Then he asked the AI about the meaning of a Zen koan. This time LaMDA side-stepped the question and its response was not nearly as smooth and focused. It had, of course, been previously fed the contents of many conversations, including ones that touched on Victor Hugo's masterpiece but not with ones containing information needed for chatting about koans which are obtuse, calculatingly paradoxical anecdotes or verbal give-and-takes used in Zen Buddhism to promote spiritual development.

LaMDA is a fascinating AI but mimicking sentience is not emulating it. Simulating consciousness is nothing like being conscious. Google put Lemoine on unpaid leave.

But, for the purpose of this book, our point is straightforward and worth repeating: life and sentience are coterminous. Life is based on cells, biomolecular entities of immense complexity. The *living* element, the biological component, is the critical factor. As Tufts University biologist Michael Levin has pointed out, computing devices are made up of inert, individual parts, none of which has any existential role to play in carrying out computations. It is not until the device is assembled that it is able to operate as designed. With

[5] For a sense of how LaMDA functions and how the database was created, go here: https://blog.google/technology/ai/lamda/

biological entities, each part, each cell, and each element within each cell has a role and a discrete function that can be carried out independently of the Gestalt, the whole. Levin's observation is important because it identifies a critical element—synthetic devices and biological ones are built on wholly different platforms.

Also keep in mind that conscious entities are error-prone, express doubt at an internal self-referential level, and appraise information using analogue processes. What is almost never addressed is that humans, above all others, are not entirely rational. We do clearly stupid things, sometimes out of ignorance or misguided thought, and sometimes for fun. We do it to attain an internal self-defined, preferential state. How might that be programmed?

Our guess, which we suspect Levin would agree with, is that if an AI is ever to become aware, have internal, valenced states graced with a sense of self and self-awareness, it will be when researchers in the future will have worked out how to integrate cellular biocomputational processes and functions into their device. This is not on anyone's horizon right now, though some (see Schumann & Pancerz, 2016) are speculating about the possibility and others (Kagan et al., 2022) are making efforts to integrate living cells into their systems as a way of achieving synthetic consciousness from the boosting platform of pre-existing sentience. To date, all functioning AIs are built on digital, informational platforms while all living organisms are formed on analogue, structural scaffolds supported by digital nucleotide sequences of RNA and DNA macromolecules. The difference is more important than many appreciate and is pursued in more detail in Chapter 8.

The issue of a sentient AI, an artefact that displays human cognition and emotion, is one of the more vigorously discussed and debated issues among philosophers, neuroscientists, computationalists, computer scientists, mystics, and futurists (see Hunt, 1995, for an overview). Not surprisingly, the debate has a long history crafted by such intellectual luminaries as the elaborately monikered alchemist, physician, and philosopher Philippus Aureolus Theophrastus Bombastus von Hohenheim (aka Paracelsus, 1493–1541) who proposed that an artificial human could be created out of magnetism and sperm using basic principles of alchemy. There is a statement in Wikipedia[6] that Paracelsus claimed he had actually done this but we could find no evidence for anything other than the speculation. Rabbi Judah Loew ben Bezalel (1528?—1609), a revered Bohemian philosopher and mystic, offered a similar proposal—only his preferred ingredients were clay and an appropriate

[6] https://en.wikipedia.org/wiki/Timeline_of_artificial_intelligence#CITEREFMcCorduck

Talmudic incantation, which he claimed to have used to create the Golem. Mary Shelly's Frankenstein monster was brought to life with electric sparks, then thought to have a role to play in the emergence of life. These early efforts were, of course, based on biological principles. The digital age of electronics and computers was centuries in the future. In an odd way, we celebrate these clever, early speculations as they did seem to grasp one of our fundamental points: life and its attendant cognitive, sentient functions are based on biophysical and biomolecular processes.

Might It All Just Be Computational?

The notion of life and mental states emerging in an artificial entity gathered steam and adherents beginning in the early 1960s, stimulated by a proposal by Harvard philosopher Hilary Putnam that the key element in creating consciousness, experience, and valenced perceptions was *computation*. If, Putnam maintained, an artificial entity were to carry out the identical operations of a human brain, its causal properties and, hence, its internal experiences, would be identical to ours. In short, you'd have a conscious robot with an existential mind. The suggestion caught on with a number of philosophers and computer scientists and led to an enthusiastic programme of research into what became known as 'hardware-independent functionalism', 'computationalism', and several other terms all of which focused on the claim that the functions were the key not the device carrying them out. As philosopher Jerry Fodor, an early and enthusiastic promoter of computationalism, was fond of saying, the system could be made out of string and tin cans but if it carried out the proper causal computations it would be conscious. It is worth noting that Fodor was Putnam's student.

In addition to enthusiastic defences of Putnam's suggestion, there were vigorous critiques of it, most notably by philosopher John Searle. Searle's arguments were based on a thought experiment, known as the Chinese Room. There is a fuller discussion of it in Reber (2019) but we can lay out the argument in simple terms. Inside the room is a person, we will call her Karen, who knows no Chinese, cannot read it, speak it, or understand it. Karen is, however, equipped with a set of instructions that can be used to decode written Chinese and craft written responses. Outside the room are several Chinese speakers who ask her questions written in, of course, Chinese. When one arrives in her inbox, Karen consults her code books for how to handle the material and how to respond to it in Chinese. She dutifully returns the replies to the folks outside the room who conclude that, clearly, whoever is in the room, is fluent in Chinese. Searle maintained that this setting is a critique of Putnam's initial argument since all the appropriate computations for reading and writing Chinese are being carried out but there is no comprehension, no meaning. Karen is acting like a chess-playing robot that can play the game at grandmaster levels without knowing anything about chess—in fact, without knowing anything at all.

Searle's argument was subjected to extensive criticism[7] and more than a few philosophers have concluded that not only is it wrong, it has been thoroughly debunked (see, e.g. Hauser, 1997). While there may be problems with the Chinese Room argument, it does not mean that computationalism is correct. Searle's argument was designed to show that the proposal for a sentient AI was logically incoherent, that syntax (what Karen's code books were based on) did not get you semantics (the meanings the Chinese speakers outside the room drew from her responses). We note that Searle, whatever the merits of the Chinese Room Gedankenexperiment are, came to the issue as a philosopher, not as a computer scientist, not as a social scientist, and certainly not as a cell biologist. He did not focus on the biomolecular, cellular elements. We agree with Searle's conclusions about functionalism and computationalism but for different reasons. Our claim is that the device carrying out the computations is not only relevant, it is *the* critical component. The device must be alive.

We will also note, in passing, that the functionalists got away with a rhetorical gambit that made resolution of the debate not only difficult, but nearly impossible. The critics, by attacking Searle's arguments, succeeded in putting those taking the anti-computationalist line on the defensive while trying to counter the anti-Chinese Room criticism. What was left in the dust was that the burden of proof properly belonged on the hardware-independent functionalists to show why their position was coherent—not on the critics of computationalism, whether they took a Searlean line or the biomolecular one we take here. Science fiction writers, futurists, film makers, and poets can all conjure up robots and AIs that behave as conscious entities but that does not mean they are possible in the real world.

Interestingly, there have been a number of recent articles and books critiquing the hardware-independence view. A common theme in the criticism is that the existing AIs, ones that can carry out many tasks that appear to be examples of intelligence including complex undertakings such as identifying

[7] The original paper (Searle, 1980) was published in *Behavioral and Brains Sciences*, an open-commentary journal, and became the journal's most influential article. Stevan Harnad, who was then the editor, wrote an overview of the debate as it played out over many years with a variety of proffered arguments both pro and con (Harnad, 2001).

the principles of protein folding, making decisions about appropriate individualized medical care, creating aesthetically appealing 'paintings', predicting the weather, and dealing with the complex problems of climate change are all domain-specific.[8] As Epstein (2016) and others point out, none of them display 'general intelligence' in the sense that the device can function effectively in arbitrary environments. Note the similarity here with Google's LaMDA and ChatGPT. The feature the programmers are working towards is to build an AI that can hold coherent conversations on any arbitrary topic, provided it has sufficient information garnered from billions of real-world conversations. And that, of course, is what humans do. If you know a lot about the cognitive neurosciences you can have conversations with other knowledgeable people, but if you're a concert pianist the back and forth is likely to be like the one Lemoine had with LaMDA when the topic was a Buddhist koan. It is not going anywhere.

Epstein does a rapid voyage through the history of brain metaphors from the early theological ones to suggestions from Rene Descartes, George Miller, John von Neumann, and Ray Kurzweil all of whom put forward one or another version of the broader claim that 'the brain is just a ____'—where the blank got filled out by whatever the insights into human cognition and neurophysiology were currently the focus of research and theory. All are dedicated to the specific task at hand while ducking the key issue: human consciousness, intelligence, is domain general. An AI can play poker and, interestingly, the latest ones do so at world-class levels. The program *Pluribus*, developed at Carnegie Mellon University, consistently wins against up to five opponents, even when all are professional poker players. But it cannot play chess. Humans can do both.

Epstein's 2016 article was intriguingly titled 'The empty brain' and his main point was straightforward: brains are not computers. The human brain does not compute in any way that resembles what digital devices do. For one, it is not digital. It is analogue in nature and will not be simulated on a digital device. In addition, living agents are sensitive to the limitations and ambiguities inherent within sources of available information (Miller, 2018). Further, as we will detail in later chapters, a cell's information is a product of its own internal analyses and is inherently ambiguous and subjective while computer data are

[8] Searle argued that these kinds of operations were examples of 'weak AI'. In his framework, they were based entirely on computations and could solve real-world problems but without emulating consciousness—which he called the goal of 'strong AI'. As he noted, one could program a computer with the code that captured everything known about a swimming pool but no one would get wet. The 'weak AI' issue would be resolved but the 'strong' one would still be untouched.

digital and defined by precision. There is no internal self-awareness of the flow of information within artificial entities.

Moreover, AIs are not embodied. Brains are based on cells, which neuroscientist and Nobelist Ramón y Cajal discovered well over a century ago (1894). They live in bodies and are ineluctably intertwined with their biomolecular functions. Bodies also move, and encounter environments in flux. Brains provide a central focus for our emotions, feelings, fears, loves, attractions, repulsions, and, importantly, function in complex social contexts and cultures. Brains also develop over time and change with experience and, of course, they age. None of these experiences or processes are, or so we maintain, possible on non-biological platforms as currently constituted. As Erik Larson noted in his recent critique of contemporary views of AI, 'all evidence suggests that human and machine intelligence are radically different' (Larson, 2021, p. 1). We, of course, are extending Larson's argument to all biological entities. The sentient functions of individual cells, as will become apparent in the material we present in the rest of the book, are 'radically different' from any algorithm-based computational system that is instantiated on a digital device.

Although not typically raised in this context, the issue of free will[9] also plays a part in our thinking and, like the other perspectives put forward, argues against the possibility of sentient entities that are not built on a biomolecular platform. Our sense of how we live our lives invites the notion that we have free will since we can, and do, make individual, contingent decisions that are exclusively our own, as they are based on our obligatory self-generation of all available information. However, all of biology is an exquisite balance of freedoms of action and countervailing constraints that manifest themselves at all levels, from protein configurations and thermodynamic constraints onward to the choices we *Homo sapiens* routinely make. Through reiteration, we manifest all of this as our own mixture of available free will and environmental constraint. Placing free will into the narrower and more direct narrative of whether or not we are capable of self-directed choices as opposed to living automatons endowed with an illusion of self-determination, the answer in our approach is quite plain. Being is doubt. We live within obligatory ambiguity. Robots do not.

[9] Our use of 'free will' differs from the way the term is typically used in philosophy, sociology, and cognitive psychology. Our meaning designates processes in cells where decisions are made based on events at the membrane and how they are interpreted in the cytoskeleton. We discuss these in later chapters. The late, highly respected Harvard cognitive psychologist Dan Wegner argued in *The Illusion of Conscious Will* (Wegner, 2002) that we don't really have 'free' will. We have 'constrained' will. We chose from a limited number of available actions and the limits on them are quite firm as they are the result of culture, upbringing, socialization, the current circumstances, etc. We agree—noting that so do cells.

In addition, as sociologist Harry Collins has argued in several places (see Collins, 2018, 2022), a critical feature of AI research is that it is based on a model that fails to appreciate that entities that have mental states, consciousness, have had extensive interactions with others—they have been *socialized*. An AI, no matter how sophisticated its computational power, would lack the 'knowing' interactions with others that would allow it to function effectively in the real world. It would not be, as he put it, 'capable of absorbing knowledge from [its] social surroundings'. Collins, amusingly, titled his recent book *Artifictional Intelligence*.

Recently, a novel approach to computationalism has been put forward by Swedish mathematician and computer scientist Gordana Dodig-Crnkovic (2022). Unlike the hardware-independent stance championed by Putnam, she argues, as do we, that the basis for sentience is biological and sees naturally occurring objects in terms of information. Essentially, her argument is that the platform for sentience must be biological info-computational and viewed as self-referential measurement of information. Her efforts are directed at unpacking the computational details that cells carry out and identifying the biomolecular mechanisms responsible. We discuss her analyses of these computational elements in cellular sentience in more detail in Chapter 9.

Overview

So, to summarize our position, which will be the framework for the remainder of this volume:

- Life and sentience are coterminous. Unicellular prokaryotes (archaea and bacteria), the original life forms and their descendants, are sentient.
- All living organisms exist in bodies composed of cells. Their functions and behaviours are intimately linked with those biomolecular, cellular mechanisms.
- These first unicellular species were spectacularly successful and provide the foundation upon which all life forms, extant and extinct, evolved.
- A basic principle of biological evolution is that a trait, structure, or function that has significant adaptive elements is virtually never abandoned or jettisoned but becomes the platform for further evolution.
- In the absence of evidence that any species or clade ever lost its sentient, affective, and/or social features, it is reasonable to conclude that cognitive functioning and sociality are essential features of all embodied life, including plants.

- The hardware-independent functionalist's core claim is fatally flawed. An AI, even one able to carry out the full panoply of cognitive functions of a species, would not have that species' sentience, internal perceptions, consciousness, or experience of embodiment. It would not feel. It would be as dumb as a rock—acknowledging that rocks can play important roles but none of them involve sentience.

When we begin the elaboration of the arguments we will be making in this book, the core presumption will be that minds are caused by and only by the biomolecular functions of living entities, ones based on sentient cells as the quintessential foundation for all life. Lastly here, we note with amusement that Putnam, who started this whole gallimaufry, later changed his mind arguing that, in fact, hardware-independent functionalism was a dead-end and a sentient computing device is not possible (Putnam, 1999).

Contents

Abbreviations

AI	artificial intelligence
ADP	adenosine diphosphate
ATP	adenosine triphosphate
CBC	Cellular Basis of Consciousness
DNA	deoxyribonucleic acid
EI*	effective information
FUCA	first universal cellular ancestor
GABA	gamma-aminobutyric acid
ITT	integrated information theory
LaMDA	Language Models for Dialog Applications
LECA	last eukaryotic common ancestor
LUCA	last universal common ancestor
PIF	pervasive information field
RNA	ribonucleic acid
ROS	reactive oxygen species
TE	transposable element

1
The Cellular Basis of Consciousness (CBC)

The CBC Theory

As noted in the Prologue, our message in this volume is built on a rather simple but surprisingly fruitful assumption: all life is sentient. Life and consciousness are coterminous. All organisms have functioning cognitive systems, from the simplest unicellular prokaryotes and protists to us, *Homo sapiens*. The model was dubbed the Cellular Basis of Consciousness (CBC). Details on how and when the theory was developed are in Reber (2019). We will be using the abbreviation 'CBC' throughout the book.

In a recent paper (Baluška et al., 2022b), we pointed out that in science the optimal strategy is almost always to begin the explorations with simple systems and only, over time, as understanding develops, move to more complex ones. The so-called natural sciences (physics, chemistry, geology, astronomy) followed this course. Unfortunately, biology and the life and social sciences did not—but for perfectly understandable reasons. They began with the most complex of systems, *Homo sapiens*, and only later looked at simpler, less complicated domains, eventually reaching down to the prokaryotes and unicellular eukaryotes.[1] One of the themes of this volume will be to explore the problems that this ordering of empirical and theoretical priorities created and examine how taking our cellular-based model can resolve many of them.

Before embarking on the voyage, let us be clear about how we are using the term 'assumption'. It is not like 'axiom', where one uses logical deduction while attempting to lay out the framework for a formal theory. Rather, we are using it in a manner similar to how Darwin introduced the principle of natural selection, Wegener began his explorations of continental drift, Metchnikoff developed his model of cellular immunity,[2] and Pasteur delved into germ theory

[1] There are good reasons for treating the earliest appearing eukaryotes as multicellular although the outward appearance of many species such as amoebas invited the initial assessment that they were unicellular. This issue is discussed in Chapter 4 where we explore the evolution of eukaryotic life.

[2] Russian–French biologist Ilya Metchnikoff's (1845–1916) work on phagocytosis led to the modern science of immunology. In 1908 he was awarded the Nobel Prize in Physiology or Medicine for it. We will discuss the pivotal role of immunological competence and its relationship to consciousness in later chapters.

The Sentient Cell. Arthur S. Reber, František Baluška and William B. Miller Jr., Oxford University Press.
 DOI: 10.1093/oso/9780198873211.003.0001

and vaccinations. In these, and many other scientific adventures, the theorist simply assumes that some proposition is true and initiates the investigations using time-honoured procedures such as examining the database, looking for converging lines of evidence, digging into the kinds of entailments the assumption suggests, running experiments, and engaging in discussions and, invariably, disputes with others who may have a different perspective on the issues at hand. And that is what the rest of this book will be about. We will look at the theory itself and, importantly, outline its various entailments. As we will see, the assumption that all life is sentient life and that the building blocks are individual conscious cells has far-reaching implications.

It will also put biology and the life and social sciences on a novel footing where the framework within which research is carried out necessarily changes and the current format no longer applies. As we will argue in several places, there are good reasons to suspect that the conceptual framework within which the life sciences currently operate has reached a critical juncture. The explanatory power of contemporary models and theories has reached its endpoint, or is very close to it. Novel approaches, new theoretical frameworks, are needed and it is almost certainly the case that they are going to have to be interdisciplinary. Israeli chemist Addy Pross maintains that a mature biology is going to have to be integrated into a larger conceptual structure that incorporates physical and organic chemistry. In his important 2016 book, he asks the question that lies at the nexus of these disciplines: 'When and how did chemistry produce biology?' What is needed is a new scientific agenda, one that focuses on getting a deeper look at the events that took place at the birth of life in the prebiotic soup—that point when physics and chemistry 'produced' biology, where inert, non-living molecules combined and coordinated their actions and living, sentient cells emerged.

The Standard Model in the Field of Consciousness Studies

We are under no illusions or delusions. We understand that the CBC model is an outlier in the broad field of consciousness studies—and it is broad, encompassing the biophysical sciences, cognitive psychology, cognitive neurosciences, and philosophy, in particular the philosophy of mind. The generally accepted stance, the Standard Model of Consciousness, is that consciousness,

It is important in appreciating the subtle but critical ways in which self-referential cells handle information and make decisions.

sentience, is a property of biomolecularly complex eukaryotic species—those that have evolved a nervous system. In this approach, the presumption is that unicellular species are not sentient and that their complex behaviours[3] are the result of automatic, genetically encoded, biochemical processes that spin themselves out devoid of feelings, of consciousness, or even valenced perceptions or preferences. In philosopher Daniel Dennett's trenchant phrase, there is 'competence without comprehension'. The classical argument is that, at some point in evolution, a sufficiently complex species or clade emerged that possessed self-referencing functions and was aware of itself and its surrounds. Variations on this approach to the origins of conscious minds can be found in one form or another in a variety of recent books and papers including those by Colin Barron and Andrew Klein (2016), Antonio Damasio (2022), Daniel Dennett (2017), Todd Feinberg and Jon Mallatt (2013), Simona Ginsburg and Eva Jablonka (2019), Tam Hunt and Jonathan Schooler (2019), Tam Hunt, Marissa Ericson, and Jonathan Schooler (2022), Joseph LeDoux (2019), and Ogi Ogas and Sai Gaddam (2022).

All assume that a nervous system is a prerequisite for an existential consciousness—although none present a coherent argument why this should be the case. They all do, however, have extensive discussions of how having a nervous system allows for a variety of new functions to emerge. LeDoux, in particular, presents extensive analyses of the neurological mechanisms that modulate and control cognitive functions in a variety of species, in particular humans.[4] It is, of course, true that once nervous systems and brains evolved they were accompanied by dramatic changes in behaviours but, and this is important, there have been a large number of functions that evolved and dramatically altered life on the planet. Multicellularity, warm-blooded metabolism, respiration, photosynthesis, a vertebra, and legs come to mind and there are many others. But each built on earlier, existing biological platforms. We have yet to encounter a solid argument that supports the proposition that consciousness is dependent on an organism having a nervous system. A nervous system changed how consciousness was instantiated but it did not create it.

[3] Which, as outlined in Chapter 2, include a variety of cognitive processes and functions such as valenced perception, feeling, developing preferences, associative learning, pattern learning, forming stable memories, decision-making, directed locomotion, social interaction, and meaningful communication. The range of behaviours that prokaryotes routinely carry out is remarkable and far more sophisticated than is generally recognized. The biomolecular structures and processes of these unicellular species are also far more complex and rich than most appreciate. These will be examined later, in Chapters 4, 5, and 6.

[4] LeDoux's book was the subject of a special issue of *Philosophical Psychology* to which two of us contributed a commentary (Reber & Baluška, 2022). Not surprisingly, we focused on the CBC and how, in our opinion, LeDoux's arguments would actually be strengthened by recognizing that all cells are sentient.

Worse, as noted elsewhere (Reber, 2019), the current standard model is saddled with an *emergentist's dilemma*—a problem that, surprisingly, none of the authors of the cited books seemed to recognize. If the claim is that a particular species, in virtue of evolving some trait or feature like a nervous system, suddenly became conscious while the ones from which it evolved were not, there is an obligation to at least take a shot at speculating what biomechanism(s) accompanied this novel transformation. We have yet to see any effort on the part of those who argue that a nervous system is a necessary and sufficient feature for a species to be conscious to explain how this miraculous shift took place. We appreciate, of course, that we have our own 'emergentist's dilemma'—but it is more tractable than the others. It is surely going to be a lot easier to identify the mechanism(s) that make prokaryotes sentient than those that emerged in insects (Barron & Klein, 2016), crustaceans (Birch et al., 2021), corvids (Nieder et al., 2020), and cephalopods (Godfrey-Smith, 2016), or why an 'unlimited associative learning' mechanism (Ginsburg & Jablonka, 2019), or a cluster of modules based on neural nets (Ogas & Gaddam, 2022) should do the trick. We've made some recent excursions into this domain (Baluška et al., 2021b, 2022b) and will dig deeper into the issue in later chapters. Spoiler alert: the candidate mechanisms currently under consideration involve processes that take place in the cell's cytoplasm and the excitable cell membranes which, for those who have not explored this domain of cell biology, are remarkably complex, intricate structures with a truly stunning array of functions.

These other avenues became the industry standard for a rather simple reason. When philosophers and scientists began explorations of consciousness, the focus of their attention was us, *Homo sapiens*. We knew we were conscious. We understood that it was a distinctly complex cluster of mental experiences, functions, and processes. There was no need for speculation or existence proofs. All we had to do was to stop and think and voila, there it was—consciousness, self-hood, a mind, precisely as Descartes famously quipped. This initial focus on human sentience had obvious consequences, among the most important was that it laid out reasonable guidelines for how to approach the topic and identified the protocols to employ. Begin by identifying behaviours that are diagnostic of human consciousness and uncovering the underlying neural structures and pathways that modulate and control them. Once there is a decent understanding of both, examine the evolutionary tree for homologues and analogues of the behaviours and underlying neurocognitive structures. As noted, this research strategy has been employed for decades and has led to a crazy-quilt array of conclusions about when in speciation consciousness first appeared. They all, perhaps not surprisingly, contradict each other.

Human consciousness, the *Homo sapient* mind, has been viewed as the pinnacle of cerebral functioning—and rightly so. No other species even gets close to our capacity for creative thought, introspection, ruminative emotions, self-reflection, language, art, and the scores of other sentient talents we have. In more recent approaches, where the focus has been on the neurocognitive and philosophical issues, the most asked questions have been ones like 'How does the brain make the mind?' and 'How can mere biomolecular "stuff" create thoughts, feelings, ideas?' This question has been immortalized by philosopher David Chalmers who put it forward as the 'Hard Problem' of consciousness (Chalmers, 1996). Chalmers, unhappily from our materialist, reductionist perspective, decided there was no possible answer to be had in what we will call 'ordinary' biochemical frameworks. He retreated into dualism, assuming that the mental was just made up of other kinds of stuff and should be dealt with separately from effects caused by ordinary matter. We do not condone this move but understand how taking a philosophical approach to the matter of consciousness could lead there. In his latest work, Chalmers dives provocatively into virtual reality—a topic far removed from our interests here but others are welcome to pursue it. Whether you agree with him or not, Chalmers' work is uniformly creative and thoughtful.

Some Views Almost in Line With the CBC

There are others who have staked out positions with regard to unicellular life that are closer to ours—some in the past such as William James, others more recently. James, in his justifiably famous 1890 work *Principles of Psychology*, opined that a conscious cell was the only position that was not self-contradictory because what cells were able to do just seemed to reflect sentience. However, after struggling with this proposition, he eventually abandoned it on the grounds that if each cell were conscious, multicellular organisms, like us, would have a mosaic of individual sensations and feelings—and that did not fit with our introspections, our sense of self which was singular. We will deal later, in Chapter 4, with this, the so-called binding problem. But, importantly, as British researcher Jonathan Edwards noted, there is an even more basic problem within the one James was concerned with. Edwards pointed out, correctly, that the binding issue was present in the actions of individual cells. As he put it, 'a bound conscious experience is a property of an individual cell, not a group of cells. Since it is unlikely that one specific neurone is conscious, it is suggested that every neurone has a version of our consciousness, or at least some form of sentience' (Edwards, 2005). We read that statement as a willingness to go along with the basic premise of the CBC.

There are coherent models that provide tantalizing insights into how multicellularity evolved and how eukaryotic species maintain a singular sentience, a coherent identity, and do not experience a medley of conflicting internal representations. We have touched on this issue elsewhere (Reber, 2019; Miller et al., 2020b; Baluška et al., 2021b) and will address it in more detail in Chapter 4. To anticipate the discussion, recognize that each cell in a multicellular organism still has its own experiences and sentience but it is *its* sentience. It is not shared with other cells. Rather, each cell responds appropriately to events and communicates with other cells. It is the assembly of all the cells involved in the circumstances that is captured and represented as the experience of the individual. James's concerns about a jumble of experiences contributed by each cell is not a real problem.

In his 2009 book *Wetware: A Computer in Every Cell*, British biologist Dennis Bray toyed with the notion that individual unicellular species such as amoebas display behaviours and processes that appear to be reflective of internal, felt states such as a sense of nutrient deficit (hunger), a need to explore the environment (curiosity), and a motivated reduction in metabolic functioning (fatigue). He also emphasized the extreme complexity of individual cells and living unicellular organisms. But he demurred on the key issue of cellular sentience: 'single cells are not sentient or in the same way that we are'. The 'same way' hedge is, of course, where the problem lies. In the CBC model, sentience is viewed as a continuum so, yes, there will be countless shifts in the 'ways' in which each species experiences their individual conscious states. But we view them, as noted, not as individual 'types' but as 'tokens' of an underlying type. The issue may have more to do with terminology than with science. It does not feel like a stretch to read Bray as supportive of the core assumption of the CBC.[5]

We've seen this kind of lexicographical confusion in other contexts. Neuroscientist Antonio Damasio recently acknowledged that the descriptions of prokaryote behaviour outlined in Reber (2019) and the follow-up discussions of even more sophisticated functions (Reber & Baluška, 2021) are suggestive of sentience (Damasio, 2022). But he criticized the use of the term 'consciousness' for them—which makes the different perspectives here feel

[5] He has confirmed this shift in his thinking (personal communication).

more like an issue of terminology and not one where there are disputes over basic biomolecular processes. The term 'consciousness', is, in the general literature in the cognitive neurosciences and philosophy of mind, so tightly linked with human mental states that it is difficult to pry it away from being semantically tied to *Homo sapient* cognition. This lexicographic tie is unhelpful as it tends to make researchers and philosophers sceptical of the CBC. They acknowledge the data for prokaryote cognition but do not want to call it 'cognition'. They agree with the remarkable functions of bacteria that are strongly suggestive of sentience but do not want to say they have an existential 'consciousness'. They appreciate the evolutionary biological arguments of the CBC but shy away from concluding that all life is sentient life. It makes our goal of changing the framework within which the field of consciousness science operates a lot harder to do. These lexicographic confounds are reviewed in Appendix I.

A recent stance is that taken by South Asian researchers Contzen Pereira and J. Shashi Kiran Reddy who have argued that indeed, all life forms are sentient, including unicellular species. They identify quantum mechanical processes as responsible and argue that the operations in the cytoskeleton that cause sentience are based on what they call 'quantum consciousness'. Their approach is reflective of the position of British mathematician Roger Penrose and American anaesthesiologist Stuart Hameroff who have put forward a theory based on quantum mechanics and space-time geometry—one that identifies biophysical functions within the microtubules of individual cells as the processes that create consciousness (see Hameroff & Penrose, 2014, for an overview). Pereira and Reddy's work has a distinctly mysterian element to it as they link conscious cells with the existence of a soul, the utilization of 'cosmic' energy, and integrate their position into a fully panpsychic framework. For details see Pereira (2015), Reddy and Pereira (2016), and Reddy (2017).

Lynn Margulis famously penned a paper in 2001 titled 'The conscious cell' in which she made the case that the unicellular organisms that combined, via symbiogenesis, to produce the first eukaryotic species were already conscious entities. In this paper she also noted that 'Nothing . . . has ever been lost without a trace in evolution'. Indeed. As noted earlier, this is a fundamental principle of evolution and we will discuss it in more detail as it forms one of the keys that gives the CBC theory much of its considerable explanatory power.

Vic Norris, at the University de Rouen (France), specifically used the term 'qualia' to refer to the subjective experiences that he assumes are a fundamental aspect of prokaryote life (Norris, 2021). As the term 'qualia' is typically used in the philosophy of mind to refer to the basic sensations of human mental life, Norris's position reflects the key principle that there is continuity in mental life. Like Margulis, he notes that nothing of significance to life is lost in evolution.

A few years back, Pamela Lyon, an American–Australian natural philosopher, reviewed the evidence for sentience in prokaryotes. Amusingly, she self-consciously (personal communication) 'cribbed' the title of the paper from Margulis. Instead of the 'conscious' cell, she called it the 'cognitive' cell (Lyon, 2015). The paper is a thorough review of the research on how bacteria adapt to the constantly shifting environments they evolved in and still inhabit. She concluded that bacteria have what she called a large and effective 'cognitive toolkit' that enables them to navigate their complex, ever-changing world. She also noted, importantly in our view, two disconnects between cell biologists and cognitive neuroscientists. One is a lack of appreciation of the range of functions in unicellular species on the part of cognitive neuroscientists who, because they focus on species with nervous systems, often fail to appreciate how sophisticated the life of prokaryotes is and tend to downplay the richness and complexity of unicellular cognitive processes. The other is a tendency to ignore evolutionary biology and, as a result, miss the point that these sentient functions form the foundation for the cognitive processes of more complex species.

We agree. We have had experiences that reflect these disconnects. We have all made formal presentations to groups of cognitive scientists and neuroscientists. We invariably encounter resistance to the CBC model, to our use of basic principles of evolutionary biology as a general framework, and to the empirical evidence itself. Despite the large and growing database that supports prokaryote sentience and despite the continuity of form and function that underlies evolution, the evidence is either doubted or interpreted as mere automatic behaviours that occur without underlying conscious experience or feeling. One of us has also presented the CBC model to philosophers. It was not pretty. There was uniform derision and outright hostility to the very notion that a cell could have qualia, experiences, preferences, and valenced perceptions.[6]

Brian Tomasik, currently associated with the Center on Long-Term Risk (London, UK), compiled an overview of the various suggestions in the literature that point to unicellular consciousness. The list of individual researchers

[6] Also, adding a balancing note here, in other talks we have given to groups of laypersons without backgrounds in biology or philosophy (Humanities Department Colloquia, Elder Colleges, Fulbright Programs) the most frequent response was a nod, a tilt of the head, and a version of 'Hmm, I hadn't thought of that. But it does make sense.'

and philosophers who have either made clear that the evidence for cognition in unicellular organisms is convincing or that they are open to being persuaded is long and contains many highly respected scholars. See Tomasik (2017) for the link to his webpage.

American biochemist J. A. Shapiro published a remarkable little paper where he noted that he now understands enough about the evolutionary capabilities of bacteria to conclude that we are required to 'recognize that even the smallest cells are sentient beings' (Shapiro, 2007). More recently, in a paper titled 'All living cells are cognitive', he re-emphasized this perspective, noting 'Without that cognitive capacity, they could not obtain nutrition essential for growth, survive inevitable ecological changes, or correct accidents in the complex processes of reproduction' (Shapiro, 2021).

English–Spanish philosopher Rupert Glasgow put forward the notion of 'minimal selfhood' based on an intrinsic reflexivity that operates independently of a nervous system and which he regards as the roots of consciousness. In his 2018 book he argues that it 'makes sense' to grant an elementary consciousness to single-celled organisms such as amoebae and dinoflagellates and 'some simpler animals'. We note that the two species identified are eukaryotes and he was unclear about whether those 'simpler animals' included prokaryotes.

Biologist Michael Levin, who directs the Allen Discover Center at Tufts University (Medford, Massachusetts, USA), is clearly in line with our CBC. In a recent paper, he focused on the ways of evaluating sentience in beings of 'unfamiliar provenance and composition'. He specifically referred to the field of 'minimal cognition', where the emphasis is on the primitive versions of cognition in species like prokaryotes, protists, and plants (Levin, 2022).

There are others who take a similar stance but, as Lyon noted, they often acknowledge that, while unicellular species are capable of highly competent actions, they are not necessarily accompanied by internal valenced states, states with preferences. As noted at the outset of this section, the CBC is not the Standard Model of Consciousness, even among those whose research is easily interpreted as evidence of sentience.

The Evolutionary Principle of Stability: Keep What Works

As noted above and in the Prologue, there is an important general principle in evolutionary biology which shores up our core assumption and forms the foundation for our view of cognitive functions across the evolutionary tree:

> Traits, forms, and functions that evolve and have adaptive value, Darwinian fitness, are virtually never entirely lost.[7] Rather, they become the basis for other traits, forms, and functions to evolve from them in what's best thought of as *evolutionary creativity*. (Jacob, 1977; Miller, 2016a)

As this process accelerates over geological time, the foundational forms become ever more firmly set in the functions and processes of every species. In short, consciousness, sentience, is not only an inherent feature of every species, it is as deeply encoded in the species' genetic make-up as the biomolecular processes that breathed life into it.

An intriguing aspect of this principle is that plants should be sentient. The flora of the planet all evolved from an endosymbiotic event involving unicellular organisms. In an earlier work, one of us contemplated this possibility (Reber, 2019) but, after a quick overview of the issue, remained agnostic. In the years since and after a more careful examination of the empirical and theoretic literature, the proper outcome was reached and, like any good scientist, a mind was changed. We will present a thorough overview of the issue of plant sentience in Chapter 11 and conclude that, however unlikely it may appear to those who have not seen the evidence, plants are, indeed, sentient. We will also discuss, as dispassionately as possible, the various ethical concerns that such a conclusion raises in Chapter 12.

An effective heuristic is to think of evolution as forming a metaphoric pyramid where the early-appearing traits are integral parts of the base and later emerging ones build on them. The ones that make up the base tend to be stable; their biochemical and genetic functions support those that evolved later. Continuity is the theme as the newer characteristics emerge, not *de novo*, but formed using those traits and characteristics from the 'layers' that evolved earlier. The later-emerging traits typically show greater variability and

[7] In the rare cases where a trait is jettisoned, it is because the loss was adaptive. As Nishimura et al. (2022) recently showed, shedding the vocal fold membranes present in all non-human primates simplified the anatomy of the vocal system and, critically, reduced the chaotic irregularities typical in the sounds produced by other primates. The simplified vocal apparatus allowed humans to produce the full phonological spectrum that led to the evolution of language.

larger differences among individuals, species, and clades.[8] The obvious implications are straightforward: (a) sentience, consciousness, should be viewed as a continuum, where each emerging variation creates a new variation on mental expression that, (b) is one that has its evolutionary foundations, its roots, in the functions and forms expressed in the species that preceded it. As noted, human consciousness thus becomes a *token* of a singular *type*—as are bonobo consciousness, raven awareness, jellyfish sentience, Venus flytrap sensitivity, and pine tree cognition. This principle is going to play a large role in the exegesis and, as we will see, differentiates the CBC from virtually all the other models of origins of consciousness that have been introduced in recent decades. See Reber and Baluška (2022) for a discussion of how approaches that focus on the discontinuities over the various species, clades, and families differ from the CBC and what the consequences of these two strategies are in terms of overall framing.

Wherever You Look

But, back to the various approaches to the emergence of minds. Because most contemporary theorists begin with human consciousness, there are, as we noted above, but two ways to establish a research programme. In one, you begin working to identify the underlying neural mechanisms that modulate and control the various ways in which consciousness is instantiated in *Homo sapiens*. Once you have a reasonable sense of what structures, pathways, and centres are responsible, you begin a backward search through the evolutionary tree looking for the most ancient species that have those neural systems, or homologues of them—the ones that could have functioned as the platform for future evolution. In the other approach, a similar analysis is carried out but with the focus on the cognitive processes and behaviours that can be argued are analogues of human sentient functions. Some of these research programmes are aimed at trying to identify the species or clades in which consciousness first evolved. Others focus on specific species or phyla in order to ascertain whether they display an existential consciousness—in these, the question of whether these species were the first genuinely conscious ones is typically not broached.

Elsewhere, we reviewed the results of these parallel programmes (Reber, 2019; Baluška & Reber, 2020). The findings have been very impressive. Scores

[8] This pyramidal model has a formal structure which we do not need to go into here but, for the curious, it is described in Shank and Wimsatt (1988) and Reber (1993).

of researchers in biology, zoology, neuroscience, psychology, anthropology, and sociology have produced a rich and compelling picture of the remarkable behaviours of a vast array of species and have uncovered an impressive range of neurological structures and functions that modulate and control them. We are strongly supportive of this research and herald the findings. But the goal, that such systematic investigation can lead to identifying conclusively the species or clade in which consciousness first emerged, has been elusive. In fact, what it has produced is a rather amusing outcome simply because the researchers and laboratories that have pursued these two methodological avenues have tended to focus on particular species. Entomologists work with insects. Avian experts with birds, in particular corvids and parrots. Marine biologists have studied a host of species including, importantly, cephalopods. Primatologists have explored the behaviours and brains of various species of monkeys and apes. What they have concluded is that the existence and/or origins of consciousness are in the species they study. From the CBC perspective, this outcome was inevitable because consciousness was always there—all that was needed was a close look.

A classic example is the paper by Nieder and colleagues (2020) who found that corvids (passerine birds like crows and ravens), despite lacking a layered cerebral cortex, behave in a visual detection task in a manner that 'is an empirical marker of avian consciousness'. Moreover, they noted that the birds' neuronal responses displayed a two-stage process that they maintained is diagnostic of cognitive processing. Their conclusion was that a 'sensory consciousness' arose independently of mammals. Nieder et al.'s findings are coordinate with the research of many others on corvids. Weir et al. (2002) reported tool-making in New Caledonian crows. Kabadayi and Osvath (2017) showed that ravens plan for the future and display impressive self-control. Müller et al. (2017) reported that ravens not only can recognize individual human faces, they can hold a grudge. Their study involved a 'bartering game' where a raven's unpalatable bread could be traded for highly preferred cheese. One experimenter cheated—took the bread but ate the cheese. Afterwards the raven would not play the game with that experimenter and refused to do so a month later. Another raven, who merely watched the episode, also refused to play the game with the 'cheating' human.

Nieder et al.'s (2020) report is a classic example of how these two lines of research are being carried out and how they fit into the overall research strategy that has dominated the field. It was published in the prestigious interdisciplinary journal *Science* and the editors considered it sufficiently important to have a separate 'blurb' on it which had this passage:

> Humans have tended to believe that we are the only species to possess certain traits, or abilities, especially with regard to cognition. Occasionally, we extend such traits to primates or other mammals—species with which we share fundamental brain similarities. Over time, more and more of these supposed pillars of human exceptionalism have fallen.

We have no issue with the last sentence. We simply repeat our core argument: any attempt to exclude consciousness from any living organism, including cells, will eventually fail.

Jonathan Birch and colleagues (2020) presented an overview of these research programmes with multiple species in an effort to answer the question 'How does consciousness vary across the animal kingdom?' They provided five 'key' dimensions of consciousness[9] that they argue are diagnostic: (a) perceptual richness, (b) evaluative richness, (c) integration at a time, (d) integration across time, and (e) self-consciousness. It is fairly obvious that they are introducing a rather subjective notion of 'richness' that we, frankly, have no idea how to ascertain or measure but it is clear that they are following the standard game plan. These are all features routinely associated with human consciousness and the scan of the evolutionary tree they provide is based on empirical evidence for each aspect of mental life using both behavioural and neurological data. But, critically, they only examined the evidence for sentience in birds, fish, cephalopods, and mammals. They find support for each of their dimensions and, not surprisingly, with varying degrees based on a host of factors. In their words:

> Our five dimensions of animal consciousness vary across and within species. Instead of thinking about variation between species in terms of levels of consciousness, we should think about multidimensional consciousness profiles.

Hunter (2021) is in agreement with this approach and, like Birch and colleagues, maintains that 'the emergence of consciousness' is slowly coming to be understood as 'research on animals yields insights into how, when, why consciousness evolved'.

Overgaard (2021) undertook a similar effort to establish conditions under which one could conclude that species other than humans could be considered to possess consciousness. His primary focus was insects arguing that, because their nervous systems were so different from mammals, identifying

[9] More recently, Birch and his colleagues expanded on their criteria and focused their explorations of animal consciousness to crustaceans. This work is discussed below.

consciousness in these species would be particularly important. We will not go through the struggles he encountered trying to make sense of the literature on the origins of minds but we would like to quote from the paper's final paragraph and note how his approach reflects the problems that accompany research carried out using the standard model. When you start with humans and work your way through the tree of evolution you are invariably going to end up here:

> In conclusion, we have no direct evidence of consciousness in insects. Furthermore, for principle reasons, we will never be able to obtain direct measures of the presence or absence of insect consciousness. ... It appears that the center of the battleground is whether specific neural substrates—observed in humans—are considered necessary for consciousness. If the answer to this question is positive, insect consciousness seems unlikely.

There are others who have been a bit bolder. Barron and Klein (2016) focused their explorations on insects, examining both behavioural and neurological evidence. They concluded, unlike Overgaard, that there is unambiguous evidence that insects have subjective experiences, what they dubbed an 'egocentric representation' of the environment. They also argued that integrated structures in the insect midbrain are 'sufficient to support the capacity for subjective experience'. They noted that neurological analyses identify the critical role of the midbrain in humans by integrating experiences of bodily representation with movement through the environment (Merker, 2005, 2007)—and that analogous structures in insect midbrain modulate similar functions.

Not surprisingly, there was immediate pushback. Brian Key and colleagues challenged Barron and Klein's neuroanatomical analysis claiming that their characterization of the insect brain was flawed (Key et al., 2016).[10] Adamo (2016) argued that Barron and Klein's analysis was not really focused on consciousness because they defined it so narrowly that the evidence they presented would also apply to robots. Adamo's conclusion was that their work was interesting but did not provide any insight into the origins of minds. But, despite Barron and Klein's efforts to find evidence for sentience in insects, they had no intentions of pushing the envelope towards the position of our CBC model. As they put it, 'consciousness also gives out somewhere. Plants do not have it. It would be surprising if jellyfish did.' We suspect they are not in support of our cellular-based approach.

[10] Key has been a central figure in the ongoing debate over whether fish feel pain. We will take a longer look at this issue in Chapter 12.

Barron and Klein's analysis also led them to conclude that consciousness first emerged during the Cambrian 'explosion', a period of extensive speciation. Others, such as Ginsburg and Jablonka (2019), agree on the time frame but for different reasons. Rather than focusing on neurological and behavioural data, they argued that the full expression of consciousness required the evolution of an 'unlimited associative learning' process, one that was capable of forming associative links between arbitrary stimuli and events and not tied to specific aspects of the environment. In their view, earlier forms of learning such as those observed in species that lack a nervous system are not diagnostic of consciousness. They refer to these earlier epistemological systems as being 'nutritive/reproductive' and assume they operate without awareness, subjectivity, or preferences.

Chittka and Wilson (2019) reviewed the literature on bees covering a rather impressive array of cognitive skills including showing foresight in honeycomb construction, the ability to learn to transport objects like small balls, and, fascinatingly, solving the ball-moving task after merely watching other bees perform it. This latter finding is particularly intriguing as the bees distinguish ball-moving by other bees from instances where no bee is present. If the bees watch the ball-moving when it was carried out by using magnets under the apparatus, the bees did not try to move it. Chittka and Wilson also reported on studies showing that bees can form visual images and anticipate upcoming events. They conclude that 'some insects qualify as conscious agents, with no less certainty than dogs or cats'.

Howard et al. (2022) reported that bees can learn parity, to distinguish between displays with an odd number of objects from those with an even number. The displays were composed of four geometric shapes and were arranged in a random fashion so that the only reliable cue was numerical parity. In addition to learning the odd versus even task, they generalized it. The original training used from one to ten objects. When tested with displays of 11 and 12 objects, the bees made appropriate categorization responses. Howard et al. do not directly confront the question of bee consciousness but they noted that their study is the first to show parity in any non-human species.

Another phylum that has attracted interest is the cephalopods, in particular octopuses and occasionally cuttlefish. Australian philosopher of science Peter Godfrey-Smith's 2016 book provides a thorough and fascinating overview of the cognitive functions of cephalopods, concluding that not only are they sentient but it is likely that this phylum is where consciousness first emerged. Recently, Ponte et al. (2022) reviewed the large literature on the issue of consciousness in invertebrates and, like Godfrey-Smith, concluded that cephalopods, in particular the common *Octopus vulgaris*, are sentient. Ponte et al. used English biologist Donald Broom's (2014) five 'essential features' of consciousness: (a) evaluation of the actions of others, (b) a memory of one's own actions and their consequences, (c) being able to assess risks and benefits, (d) some degree of awareness, and (e) experiencing affective states. Note that these features have some overlap with those identified by Birch and colleagues (above). However, unlike Birch, in Broom's view any species that displayed at least one feature should be considered sentient. In Chapter 2, where we examine the behavioural repertoire of the prokaryotes, it will become clear that these *ur*-life forms, archaea and bacteria, display *all* of Birch's and Broom's criteria for having a mind.

One more example of how this research strategy plays itself out is the recent work with crustaceans, much of it carried out by philosopher Jonathan Birch of the London School of Economics (London, UK) and collaborators (Birch et al., 2021; Crump et al., 2022). In Birch et al. (2021) they extended earlier analyses (discussed above) and developed a list of eight criteria that they felt captured what defined sentience, with a specific focus on the capacity to feel pain and stress. They reviewed over 300 published papers on the issue and concluded that crustaceans are sentient, feel pain, and show negative reactions to stress. Interestingly, the final report convinced the government in the UK to declare decapod crustaceans (shrimp, lobsters, crayfish, and hermit crabs) and cephalopod molluscs (squid, octopuses, and cuttlefish) as sentient species and to include them in the national Animal Welfare (Sentience) Bill. These species are now covered by the various regulations in the UK that are designed to minimize pain and suffering and promote animal welfare in commerce. The report did not come to any determinations with regard to the origins of sentience, only that crustaceans are.

Some 50 years ago, philosopher Thomas Nagel penned one of the most cited works in the philosophy of mind: 'What is it like to be a bat?' (Nagel, 1974). The paper was a critique of mechanism and reductionism and his answer was that we will never know—and one of the reasons was because of what is known as the 'first-person' problem, that the only mental states we can be certain of are our own. Conclusions about sentience or consciousness in others, independent of the species, is just speculation and fatally, in his assessment, cannot be either demonstrated or falsified. But independent of Nagel's larger concerns (which show up in a variety of his works), the 'what is it like' question continues to be routinely asked when other species are the topic. The standard parsing of it is that one can conclude that a species has an existential consciousness when there is compelling evidence that there is something it is like to be a member of that species. The conclusion that each of the various research programmes we have discussed here is, 'Yes, there is something it is like to be a bee, a crab, octopus, or raven.' In Chapter 2, we will present the evidence that points to the conclusion that 'There is something it is like to be a bacterium.'

Interestingly, despite these converging lines of evidence and argument, there remains a stubborn lack of agreement in the broader field of consciousness studies. An interesting example is the position of British psychologist Euan Macphail who is of the opinion that consciousness requires the capacity to develop relationships of 'aboutness' or 'intentionality'. German philosopher Franz Brentano[11] introduced this concept to refer to the ability to mentally represent objects, events, properties, and states of being in the environment, to have mental states that are 'about' that which is external to the person. In Macphail's view, an existential consciousness is built around aboutness and requires language. Only humans with language have genuine conscious states. Children are not fully conscious until they have acquired language, intentionality, and can differentiate self from non-self (Macphail, 1998). Our stance, as we will see in Chapter 2, is that the unicellular species express compelling evidence of intentionality.

Finally here, if any readers are interested in exploring these issues further, we recommend the relatively non-technical work that is routinely published in the online journal *Animal Sentience* edited by Université du Québec à Montreal psychologist, Stevan Harnad. The journal operates on an open peer commentary model. Articles are peer reviewed and, after being published, other scholars and researchers are welcome to submit commentaries, which are themselves peer reviewed. This strict vetting process gives readers assurance that the work is solid and the evidence is compelling.[12] The original authors typically reply to the various comments. Many of the issues raised in this chapter are discussed there and the exchanges can be lively. Interestingly,

[11] Brentano (1838–1917) viewed intentionality as the factor that distinguished human minds from all other objects. Only humans, in his view, could form such mental representations. He was a fascinating and influential thinker with deep concerns over moral and ethical issues. Ordained a priest, he 'defrocked' himself twice. Once over his profound disagreement with the Church over the issue of papal infallibility and again when he married. At the time, the Austro-Hungarian empire did not permit married priests to hold university professorships even if they had resigned their ordination. He was allowed to continue at the University of Würzburg, Germany, but only as a docent and not a professor.

[12] We approve of Harnad's peer-review procedures as they conform with most responsible scientific outlets. Unfortunately, in recent years the sciences have been flooded with 'junk' journals and 'predatory' ones. The 'junk' journals do not send the submissions out for review by experts but generally publish any paper sent in. The 'predatory' ones follow the same unscholarly protocols but charge outrageous fees, often several thousands of dollars (US). If a reader stumbles across an article from one of these, it is difficult to impossible to know whether the work is serious, the data real, and the findings replicable—particularly because the publishers typically give the journal prestigious-sounding names that make the reader think they are respected in the field. All three of us are routinely sent 'invitations' to publish our work in them. Our 'delete' buttons are all getting work-outs.

authors of the original target article will often identify individuals they would like to receive comments from. The Crump et al. (2022) paper cited above on the criteria for determining which species are sentient was published in *Animal Sentience* and the authors specifically asked one of us to respond. We did so in a collaborative commentary (Reber et al., 2022).

2

It's Cells All the Way Down

Introduction

The title of this chapter is stolen from an old joke that (old) philosophers like to tell—and one of us almost qualifies. It goes like this:

> A young man, searching for the answer to a question that has bothered him for a long time, climbs high into the Andes to probe the mind of an ancient shaman, known for his knowledge of all things.
>
> 'What, Sire' he asks the wizened one, 'keeps the Earth stable, fixed in the heavens? Why does it not fly wildly about in the cosmos or fall into a black nothingness?'
>
> 'Ah grasshopper', the shaman says, nodding sagely. 'It is because it rests upon the back of a great turtle.'
>
> 'Interesting. Thank you Sire. I will contemplate . . .' says the young man. He walks away, but only a few steps. He turns and says, 'I'm sorry Master, there still appears to be a problem. What keeps the turtle from simply falling into the abyss?'
>
> 'Because, my child, it sits upon the back of an even greater turtle.'
>
> 'But . . .' and that's all he gets out.
>
> The shaman narrows his eyes and says, 'Look, young whippersnapper, it's turtles all the way down.'

And so it is—but with cells. All life is built on, by, and with cells. As we argued in Chapter 1, those cells are all sentient entities with experiences, internal representational states, valenced perceptions, and feelings. The rest of this chapter is devoted to an exploration of the literature that supports this, the core message of the CBC: life and sentience are coterminous.

The Time Frame

Before exploring the evidence for the CBC, it will help to have a temporal framework for when life first appeared and evolved. The estimates vary but most peg the appearance of the last universal common ancestor (LUCA) at

The Sentient Cell. Arthur S. Reber, František Baluška and William B. Miller Jr., Oxford University Press.
 DOI: 10.1093/oso/9780198873211.003.0002

between 3.9 and 3.5 billion years ago (Betts et al., 2018)—although Papineau et al. (2022) recently presented evidence suggesting that the first cells may have appeared 4.28 billion years ago. Pushing back the timeline nearly a billion years is a significant claim. We await follow-up research. Part of the difficulty in establishing a clear timeline is that single, non-colonial cells do not leave microfossils. The evidence comes from microbial mats that had grown large enough to leave their mark in the fossil record. The very first prokaryotes were social and readily formed colonies so the evidence from the microbial mats is likely to be proximate to the first cells. This colonial feature of early prokaryotic life is important and we will look at it more closely in Chapter 3.

While there are various estimates on when the Earth's environment had settled, water had accumulated, and it had reached a biofriendly state, the most common assumption is that this occurred some 4.3 billion years ago. It took, based on the dating of the LUCA, somewhere between 1 billion and 100 million years to solve that first puzzle, bringing together the materials needed for, as biochemist Addy Pross put it, 'chemistry to produce biology'. For the next roughly 1.5 billion years all life on Earth was unicellular. The only living forms were the domains of bacteria and archaea—living along with viruses. The question of whether viruses are truly alive is a complex one. We have discussed this issue elsewhere (Miller et al., 2023) and touch on it in Chapter 3. The first eukaryote appeared some 2 billion years ago. In short, solving the multicellularity problem took a lot longer than the initial problem of cobbling together the materials for the appearance of unicellular life. The first eukaryote cell was the result of a larger prokaryote absorbing and incorporating a smaller one, a process known as endosymbiosis. There are more than a few theories about the details of this event, though recent genetic analyses suggest that at least one of the two cells was archaea (Mattiroli et al., 2017).

Various Proposed Criteria for Consciousness

As we noted in Chapter 1, several researchers have put together what they maintain are the behaviours or functions that are diagnostic of consciousness. Here are three examples:

Broom (2014):

- evaluation of the actions of others
- a memory of one's own actions and their consequences
- being able to assess risks and benefits

- some degree of awareness
- experiencing affective states.

Klein and Barron (2016):

- integration of information from sensory inputs
- access to memories
- interoception
- monitoring feedback.

Birch et al. (2020):

- perceptual experiences
- evaluative experiences
- integration of information at a point in time
- integration of information across time
- self-consciousness.

There is considerable overlap, which is appropriate since, in principle, they are all focused on the same topic. There is also the inescapable feature that these criteria were developed by researchers who began with human mental life and then worked back through simpler species trying to determine which were sentient. Klein and Barron were, as noted earlier, specifically looking for evidence of consciousness in insects. Both Broom and Birch and colleagues were examining the evidence for sentience in other species with a focus on the issue of animal welfare. But, despite these differences in approach and motivation, the one thing that is fascinating is that all of these behaviours considered to be clear markers of an existential consciousness are, in fact, found in unicellular bacteria, archaea, and eukaryotes.[1] The rest of this chapter is devoted to a review of the relevant literature.[2]

As Pamela Lyon noted, cognitive scientists and neuroscientists have largely ignored the work on prokaryotes and unicellular eukaryotes and are generally unaware of just how sophisticated their behaviours are (see her overview of prokaryotic function; Lyon, 2015). The result, as noted in Chapter 1,

[1] As noted above, the earliest eukaryotes resulted from an endosymbiotic merging of two prokaryotes. Although their physical form is that of a unicellular species, there are good reasons for treating them as multicellular (see Chapter 4 for further discussion of this oft-neglected point). However, we will include them in this chapter as they display a host of behaviours and functions that are further evidence of cognitive processes in relatively simple species.

[2] Some of which was discussed previously in Baluška and Reber (2021), Baluška et al. (2022b), and Reber (2019).

is a field dominated by a standard operating position—one that assumes that whatever it is that these 'simple' species do is the result of insentient genetic programmes that operate without feelings or internal, experiential states. This stance has led the field of consciousness studies to its current muddled state. In this chapter we review the literature on the 'conscious cell' (Margulis, 2001) and the 'cognitive cell' (Lyon, 2015) that lies at the heart of the CBC. As we will see, prokaryotes and unicellular eukaryotes are rather remarkable organisms with a wide array of behaviours, all of which are clear manifestations of sentience. They:

- have elaborate sensory systems
- have valenced perceptions
- learn a variety of tasks
- form stable memories
- anticipate upcoming events
- make decisions
- process information and form preferences
- display motivated locomotion
- have temporal awareness, bio-clocks
- communicate within and between species
- cooperate with each other
- compete with each other
- are capable of computation
- trade resources
- exhibit sociality.

Not quite the sorts of things one expects to see in insentient species.

Unicellular Sensory and Perceptual Systems

The cell membranes of bacteria and archaea are covered with literally thousands of chemoreceptors. It has been known for decades that prokaryotes are sensitive to a rich array of events in the environment including temperature gradients, light levels, electromagnetic stimuli, the passage of time, the identity of specific nutrients, pH levels, the concentration of solutions (osmolarity), other bacteria (using quorum sensing to form colonies), the presence of amino acids, waste products from other organisms, and toxic objects in the environment.

Critically, these sensory functions are valenced. The positive and negative elements are experienced and evaluated which is why we used both 'sensory' and 'perceptual' in the heading for this section. The former is generally used to refer to the detection of stimuli and events, the latter for forming interpretations of that which was sensed. Hearing a series of notes played on a violin is 'sensing'. Recognizing that they are a pleasing, familiar melody is 'perceiving'. Since prokaryotes routinely make decisions based on the positive and negative aspects of objects and events, it is clear that they have well-developed perceptual processes that are the source for making decisions and forming preferences based on what was sensed (Miller, 2013). They do not simply detect that the temperature has risen to potentially dangerous levels, they actively swim away into cooler waters—which they know are cooler because they have a memory of the prior temperature to compare it with. If they make contact with an acidic molecule, they experience it as a negative object and withdraw.

Escherichia coli, and other species of bacteria, travel by two means. Rotating their flagella allows them to move forward in a linear swimming motion and random tumbling. As Berg and Brown (1972) showed over 50 years ago, they can detect the presence of positive conditions and move forward using a 'biased random walk' and change directions when sensing that things are taking a turn for the worse. Miller et al. (2021) pointed out that these kinds of chemotaxic sensory systems are adaptive and essential for unicellular species to be able to navigate environments that are in constant flux and ambiguous in the nature of the information provided. Bruni et al. (2017) showed that *E. coli* deal with these constantly changing environments by using voltage-induced calcium flux to gain information—a mechanism that is highly similar to one used in the sensory neurons in vertebrates. Again, we refer back to the 'pyramidal' model of evolution discussed in Chapter 1. If something works, it remains part of the ongoing process of evolution and forms the basis for future traits and functions. The calcium flux mechanism has been around for a very long time.

The laboratory of Swiss biologist Alexandre Persat recently reported that, in addition to using chemosensory processes to navigate their world, at least one species of bacteria (*Pseudomonas aeruginosa*—a multidrug-resistant pathogen) also evolved a sense of touch based on mechano-tactical functions (see Kühn et al., 2021, for details). They make motility decisions by literally feeling their way across surfaces and make movement decisions based on the relative stiffness of the surface.

An amusing but effective sensory experience in bacteria is their ability to detect and respond to, what the scientists at Tufts University called the 'funky' smells of the volatile gases produced by the fungi that are responsible for ripening cheese (Cosetta et al., 2020). Their report noted that this bacterial skill is likely to have commercial value in the cheese-making world.

Cyanobacteria are sensitive to the passage of time and have circadian clocks that use bioelectric mechanisms to sense the environment and, literally, anticipate regularities in the physical environmental cues such as light and temperature and use this information to make appropriate adjustments in molecular function (Baluška & Reber, 2021b). Cyanobacteria are also photosynthetic and were critical in the evolution of plants. We will have more on the fascinating moment in evolution when a cyanobacterium merged with an alpha-proteobacterium and the first chloroplast appeared. There are reasons for thinking that this endosymbiotic event occurred exactly once in the late Precambrian era—roughly 1 billion years ago (though there are questions about the date)—and that all plants evolved from it.

There are dozens of additional reports of sensory and perceptual experiences in unicellular species but these are sufficient to make the case. Prokaryotes sense, experience, feel, and use these internal representations to drive decisions, motivate actions, and influence the formation of preferences. These mechanisms are complex and have a comparative basis to them. In addition to sensing and processing information, cells must also be tuned to their own internal, metabolic conditions. They have to be aware of a nutrient deficit to move towards a food source, and note internal temperature to find the way to comfortable temperature gradients in the surrounding environment. They need to be able to navigate a complex, changing environment in order to find the metabolic 'niche' that promotes survival. German microbiologists Tobias Schweinitzer and Christine Josenhans (2010) reviewed the literature on sensory systems in unicellular species based on processing some form of energy and found that they are so widely employed by so many different species that they termed them 'global'.

Learning: Adaptive, Avoidant, Navigational, and Pattern

All unicellular species learn. We will take a look at four examples noting that they are associative in nature. That is, the learning consists of forming links between stimuli and responses. In Chapter 1 we noted that one of the criteria for consciousness cited by Ginsburg and Jablonka (2019) was an 'unlimited associative learning'. They did not, however, grant sentient states to prokaryotes.

Adaptation

Cells must have the ability to adapt to the ongoing, chaotic, external environment. The standard picture, based on early work by Jacob and Monod (1961), had been that regulatory networks evolved to establish effective gene expression states for commonly encountered environmental changes. In short, the mechanisms were assumed to already be in place as the result of evolutionary pressures. Until recently, it was not known how cells would react when confronted with unfamiliar environments, ones that are not in their evolutionary history. Saeed Tavazoie's laboratory at Columbia University (New York, USA) explored this issue recently and found that cells can, indeed, adapt successfully to new and potentially lethal environments. In their studies, metabolic functions were modified so that cells were unable to make the changes in gene expression that would normally occur when levels of a particular ligand reached dangerously high levels. Rather than succumbing, the cells carried out stochastic adjustments in metabolic function by randomly increasing or decreasing relevant gene expression. The ones that worked, which were effective in promoting survival, were reinforced and, over time, optimal levels of gene expression were established without genetically predetermined pathways. In short, cells learn to tune metabolic functions to adapt to novel, challenging circumstances (see Freddolino et al., 2018, for details).

Yang and Tavazoie (2020) recently reported a similar finding with yeast cells (*Saccharomyces cerevisiae*) exposed to ethanol at levels that crossed what they called the 'lethality threshold'. Since simple alcohols were not present in the original environments of yeast, the normal environmental stress response would not be activated. Nevertheless, over generations, the yeast cells developed an adaptive compensatory gene expression reprogramming mechanism—one that was sufficiently successful that enabled them to not only survive, but display growth rates that were indistinguishable from control cells.

Avoidance

One of the earliest studies was carried out by American biologist H. S. Jennings who reported an instance of sophisticated avoidance learning in a sessile ciliate *Stentor roeselii* (Jennings, 1906). Jennings dropped an aversive dye into *S. roeselii*'s trumpet-shaped intake funnel. The response was a classic escape response as *S. roeselii* bent over and shook the particles out and retracted its cilia. After several additional administrations, *S. roeselii* contracted so that the dye did not enter the funnel—a classic avoidance response. After a time, *S. roeselii* uncoiled and its cilia returned to their normal functions. Jennings then reintroduced the dye—which was just too much for *S. roeselii*. It tore itself loose from its perch, swam away, and reattached elsewhere—an even more complex avoidance response.

There is an interesting history to this study. In 1967, Reynierse and Walsh reported that they were unable to replicate Jennings' findings and the existence of avoidance learning in the single-celled eukaryotes was widely doubted. It took another half-century to restore *S. roeselii*'s reputation. Dexter et al. (2019) noted that the non-replication was not a real replication for Reynierse and Walsh used a different species, *S. coeruleus*. Dexter, along with colleagues at Harvard and Dartmouth, ran a series of trials with *S. roeselii* designed to mimic Jennings' procedures and reproduced the original findings. The organism engaged in the four stages of bending, ciliary retraction, contraction, and detachment first reported by Jennings. In the same year Trinh et al. (2019) reported an additional replication. Avoidance learning in a single-celled eukaryote is now broadly accepted.

Navigational

Audrey Dussutour's research group in Toulouse reported that slime moulds (*Physarum polycephalum*) learn to navigate a pathway between two Petri dishes that had caffeine or quinine on it, substances that they find aversive. Over several days they showed significant improvement in their ability to negotiate the bridge to the dish containing nutrients without coming into contact with them (see Boisseau et al., 2016, for details). In their report they described the effect as 'habituation' which we found misleading. Habituation is a phenomenon where responses to a particular stimulus diminish over time as attention is no longer drawn towards it. It is clear that slime moulds know where they are, where they are going, and learn the safest path to take. It is worth pointing out that habituation rarely occurs with aversive stimuli. To do so would not be adaptive—the presence of stressful events and objects may be unpleasant but, evolutionarily speaking, it is adaptive to pay attention. What the data do imply is what we have called here *navigational learning* and is based, like the research with *S. roeselii*, on avoidance of aversive circumstances.

Dussutour, in a recent review of the issue of learning in unicellular species, hedged on whether there is clear evidence for learning in unicellular species (Dussutour, 2021). The issue here, we suspect, is yet again, more one of terminology than of empirical evidence. Her laboratory's titling a paper that presents evidence of learning as 'habituation' suggests that we really need to clarify the meanings of terms used in the field. See Appendix I for our lexicographic efforts.

Recently, Kramar and Alim (2021) and Cheng (2022) independently reported similar findings where *P. polychephalum* learned to navigate tubes in which food had been placed at random locations. Interestingly, both reports refer to the ability to learn and form memories of the location of nutrients 'intelligent' behaviour. Neither regarded the effects as ones of 'habituation'.

Pattern

Israeli biologist Yitzhak Pilpel's laboratory showed that both *E. coli* and a yeast (*S. cerevisiae*) can learn to 'foresee' (their term) events and anticipate their arrival by initiating metabolic changes prior to their onset. The procedure consisted of presenting a sequence of two sugars that both species are fond of, lactose and maltose, one after the other in an alternating pattern, LMLMLM (see Mitchell et al., 2009, for details). When the next cycle began with maltose, nutritive uptake was inefficient. Both species had already made adjustments in gene expression designed to maximally absorb the anticipated lactose. Interestingly, Mitchell et al. were able to extinguish the pattern of shifting gene expression by presenting an ongoing series of lactose without changing to maltose.

Similar findings where the stimulus modifications were in temperature were observed with *E. coli* by Saeed Tavazoie's laboratory. They raised the bacteria in environments where the temperature was shifted from relatively cool to warm and back again, repeatedly (details are in Tagkopoulos et al., 2008). Over several generations the bacteria learned to shift gene expression to anticipate the next temperature change.

In a recent paper, Australian neuroscientist Brett Kagan and colleagues reported that neurons in a Petri dish (what they call the 'DishBrain system') learned, in a surprisingly short period of time, to play the classic arcade game 'Pong' (Kagan et al., 2022). As they put it, 'Using the DishBrain system, we have demonstrated that a single layer of *in vitro* cortical neurons can self-organize activity to display intelligent and sentient behavior when embodied in a simulated game-world'. For the DishBrain cells, their self-organization was directed, by definition, towards states of collective preference, that is, one collective state over another. However, we need to point out that, despite their findings, Kagan and co-authors cling to the standard model and state that they regard the cells in DishBrain as being 'intelligent' but not 'conscious'. From the CBC perspective, of course, we view this to be more of a terminology issue than an existential one. Their DishBrain cells are as sentient as any.

These findings also fit with the recent discovery that xenografts taken from human neural tissue thrive on rat cortex, indicating that they sense their new environment, rapidly self-engineer required internal adaptations to participate with other neurons, associate with them through collaboration, cooperation, and trade resources. Put simply, they function as participant multicellular co-engineers, and are capable of individually contributing to the entire brain output, showing that they retain their own self-identity (Revah et al., 2022).

Memory

Let us start this section by noting that whenever you have organisms that learn, you are, necessarily, going to have ones with the capacity to establish memories of what has been acquired. Without forming stable memories, the task would have to be learned anew each time the circumstances or those close to it re-occurred. Learning without memory is clearly non-adaptive, even for the simplest of organisms. Interestingly, Robert Macnab and Daniel Koshland presented evidence for memories in bacteria over a half-century ago. But they shied away from using language that would make it seem like bacteria displayed genuine cognitive functions. They put *memory* in scare-quotes (it appeared as 'memory') or qualified it as 'memory-like' (Macnab & Koshland, 1972).

Since we have already reviewed a number of experiments showing that prokaryotes and unicellular eukaryotes learn in a variety of circumstances, it should be obvious that they have parallel memorial functions. The bacterium that learned to move out of an environment that has grown uncomfortably warm has to be able to recall the temperature of the surroundings it left in order to determine whether the one it travelled into has positive valence. If not, it will move on again. The slime moulds that learned to navigate a ramp with caffeine and quinine must be able to recall the location of the aversive acids in order to travel safely over the bridge to the dish with nutrients. With this in mind, the following sections deal with some of the details of the memory systems of unicellular species.

Time Frame

Most prokaryote memories are short term, lasting usually less than a minute, some as short as 3 or 4 seconds. Their function, as noted, is to keep track of the current state of their environment so they can make appropriate decisions with regard to changes in ambient temperature, concentrations of nutrients, and presence or absence of toxic substances. In these normal environments there would be little gain to establishing long-term memories. In other circumstances, longer-lasting memories are created, in particular those that are based on alterations in gene expression. Swiss biologists Roland Mathis and Martin Ackermann (2016) showed that a sessile bacterium, *Caulobacter crescentus*, can establish memories that last up to 2 hours which, in the lifespan of a bacterium, is a near-eternity.

Cell division in *C. crescentus* is asymmetric with the stalked 'mother' cell remaining attached to its spot while the motile 'daughter' cell swims away to find another perch. Before cell division, Mathis and Ackermann introduced an aversive, concentrated salt solution which, previous research had shown, results in shifts in gene expression as a protective measure. Then, after the daughter cells relocated, they reintroduced the toxic salt solution. Death rates were much higher among the daughter cells that swam away to a presumably safe location than in the sessile mother cells who remained. The mother cells had established a memory of earlier events and displayed a readiness to make appropriate metabolic shifts should the salt solution return. Those that relocated did not. Importantly, both groups of cells had identical genes.[3]

Immunological Memory

Bacteria and other prokaryotes routinely establish memories based on immunological functions. Invasion by a virus or a strand of nucleic acid (a 'plasmid') initiates a sequence of responses by the cell. It uses a protein, known as Cas9, to snip out a piece of the invader's genetic material and insert it into their own DNA. The resulting modification of their genome functions as a memory of the identity of the invader and the cell is better prepared for coping with any future occurrences. This procedure is, of course, the basis for CRISPR, the revolutionary gene-editing technique. Jennifer Doudna and Emmanuelle Charpentier may have won the Nobel Prize in Physiology or Medicine for

[3] The role that genes play in these adaptive learning and memory functions is complex. We discussed the issue in Baluška and Reber (2022) and present more detail on it in Chapter 11.

their brilliant work with CRISPR but the technique was invented by prokaryotes several billion years ago.

Of interest is the question of the precise biomechanisms used to establish what is, essentially, a lasting representation of events that occurred, or behaviours that were carried out. Biologists Josep Casadesus and Richard D'Ari at the University of Seville, Spain, pointed out that there are, in fact, several distinct ways to encode and hold information (Casadesus & D'Ari, 2002). Some are basically genetic in that a heritable function, encoded in DNA, is established based on experience. The result is a modification of the organism's genome to a different pattern of gene expression based on the specifics in the stimuli to which it has been exposed. The results of Pilpel's and Tavazoie's laboratories on anticipation and pattern learning discussed above are based on this biomechanism as are those that utilize the CRISPR technique. Casadesus and D'Ari described these processes as ones that 'lock in' the change as it is carried onto future generations. Other techniques for forming memories are based on modifications in cellular functions that deal with repeated events or actions and are not heritable. This knowledge dies with the organism that learned and encoded it.

Decision-Making

Many of these findings on pattern learning and memory can be traced back to the early work of French biochemist Jacques Monod carried out in the 1940s. In a series of seminal experiments, he provided *E. coli* with glucose (preferred) and lactose (nutritious but non-preferred). He observed distinct pauses in the growth of the colony at those points where all the glucose was consumed and only lactose remained. The reason for the pause, Monod discovered, turned out to be one of the most important early findings in cell biology. When glucose was abundant, *E. coli* simply turned off the gene expression functions that increased their ability to metabolize lactose and needed time to make metabolic adjustments to efficiently metabolize lactose. Literally hundreds (perhaps thousands) of studies (like those described above) sprung from this finding for which Monod was awarded the Nobel Prize.

The focus of much contemporary research is on the molecular basis that cells use to 'decide' to switch gene expression (and, yes, 'decide' is the term used in the field). Brandeis University (Waltham, Massachusetts, USA) biophysicist Jané Kondev reviewed the several theoretical gene expression models for how well each handled the data in the diet-selection studies (Kondev, 2014). He had a section with the provocative heading 'Does *E. coli* have free will?' in which he concluded:

> The result is a population of cells that, although genetically identical [to another], may behave quite differently. It is as though each *E. coli* cell is free to choose between two diets, glucose or lactose, unencumbered by its genes or the environment. Experiments . . . reveal the molecular nature of that 'free will' and trace it to the stochastic expression of *E. coli* genes.

Kondev's primary interest is mathematical modelling and, as we saw in our earlier review of the work in Saeed Tavazoie's laboratory, the notion of a stochastic 'choice' and the feedback from the one made is a powerful one. The stochastic feature allows for multiple options and the decisional distributions emerge when 'free' or unconstrained choices are made. In their approach to this issue, the circumstances that are present when the bacteria choose can be artificial or organic. It does not matter because they are looking for general mechanisms, ones which operate when the environment changes no matter what the factors of change are. Whether Kondev's 'free choice' is something that falls under the philosophical umbrella put up by Daniel Wegner is far from clear,[4] but his point is made. When an organism, even a unicellular bacterium, is able to choose among alternative modes of action, it surely falls within the folk psychology sense of 'free will'.

Communication

Prokaryotes are highly social and readily form collectives known as microbial mats and biofilms. The former are laminated sheets found mostly in aquatic settings, can be up to a centimetre thick, and are held together by sticky extracellular substances. Biofilms are only a few layers in thickness and form on surfaces, commonly on boat hulls and, of course, teeth. Their biomolecular structures and functions are well known as are the critical roles they played in evolution. But here we are concerned with their cognitive functions which include the ability of cells to communicate with each other based on what is known as *quorum sensing* or *quorum signalling*.

As a bacterial colony forms, cells interact with each other by releasing molecules that have signalling functions that provide critical information to the rest of the collective enabling them to control gene expression and respond

[4] Wegner's 2002 book *The Illusion of Conscious Will* is an unpacking of the notion of will that argues, persuasively in our minds, that we don't have anything like what the average layperson thinks of as 'free will'. What we do have is a more mundane but more understandable range of things we can choose to do, or not, under the momentary circumstances. We have, in short, the capacity for choice but it is far from 'free' and, ultimately, the notion of 'will' is best handled metaphorically. The issue of course, is far from resolved (see Baggini, 2005).

appropriately to population density.[5] When a colony like a biofilm reaches a certain size, population density issues can arise. Cells on the periphery, the outer surfaces, are in a nutrient-rich environment but are also exposed to its toxins, antibiotics, and predators. Those in the interior are in a more protected location but often are faced with lower levels of nutrients. In fact, if the number of cells on the periphery becomes too large, the inner cells could literally starve to death—which would leave the colony at risk from all sides. To prevent this catastrophic event, the cells in the interior secrete a molecule that signals to the rest of the colony their state of caloric distress. The cells on the periphery respond by adjusting metabolic functions, reducing cell division rates, and decreasing the uptake of nutrients which allows sugar molecules to flow inward. Once nutrient levels in the interior are back to normal, a different biomolecular signal is emitted that says, in essence, 'We're okay now', and the outer cells resume normal activity. When nutrient levels in the interior drop again, as they must, the phase-cycling repeats and will continue to do so for as long as the collective lives.

In the event that the biomass has grown large and is made up of different species (as it often is), the form of communication changes and the interior cells shift over to releasing potassium ions. These positively charged wave-like signals travel throughout the collective and have the same effect in communicating to the outer layers the need for them to make adjustments in function.[6]

This elaborate communication network is also seen operating between two separate bacterial colonies. Gürol Süel and colleagues at the University of California, San Diego (USA) allowed two collectives of the bacterium *Bacillus subtilis* to develop in the same Petri dish and carefully controlled the nutrient levels in their shared environment. When nutrient levels dropped in the area of one collective, the potassium release mechanism was engaged, travelled through the dish's medium, and alerted the other of their distress. The response was a primitive form of what can only be called *altruism*.[7] The well-fed colony lowered molecular functions, reduced cell division, and reduced nutrient uptake until the first collective released another molecule saying they had recovered (see Prindle et al., 2015, for details). Süel's laboratory referred

[5] In passing, it is worth noting that quorum sensing is also used by more complex species including the hive insects like bees and termites to make decisions about when a nest has exceeded its optimal size and should split.

[6] The details of these communicative processes can be found in Beagle and Lockless (2015) and more on quorum sensing is in Schertzer et al. (2009).

[7] This is an interesting issue and we will discuss it further in Chapter 12. The reason we are calling it *altruism* is because it has the two key features typically identified: (a) the response increases the well-being of others while (b) it puts the individuals making it at risk. Note that the Humphries et al. (2017) finding is a classic case of non-kin altruism.

to this as 'time sharing' and Humphries et al. (2017) observed it taking place between different species.

Süel and colleagues' work showing cooperation and 'time sharing' of nutrients is fascinating and has been replicated by other laboratories, including biologist Kevin Foster's group (Sharp & Foster, 2022) at the University of Oxford (UK). However, they have also reported competition among bacterial colonies, in particular when they are of different species (Palmer & Foster, 2022). The situation is, as their recent work shows, more complex than originally thought. Since these bacteria normally live not in Petri dishes but in hosts, it is becoming increasingly clear that the host microbiome also plays a role. Depending on how it interacts with the several bacterial species, either cooperation or competition can result (Bentley et al., 2022). Foster is optimistic about the possible medical role of the discovery of competition among different species of bacteria. It hints at the possibility of developing ways of dealing with bacterial infections without the need for antibiotics.

Temporal Awareness—Bio-Clocks

Virtually every cell has a circadian bio-clock. We view them as inherently cognitive for the simplest of reasons. They provide information that can be used to anticipate the regularities in physical environmental cues such as light and temperature. Recent discoveries have revealed their deep cellular basis. The original view of circadian clocks in humans and animals was that they are controlled by cerebral processes and needed neural systems to function. However, more recent studies revealed that organs, tissues, and individual cells are running lower-level, semi-autonomous clocks based on individual cellular temporal mechanisms that are sensitive to light levels. In fact, it is even a more complicated issue than just our own cells. Our biorhythms interlink along complex communicating, metabolic pathways between our body cells and our constituent microbiome, with significant effects on satiety, sleep cycles, and even obesity (Rácz et al., 2018).

Keep in mind that photosynthesis, the basis for redox-based circadian clocks, was 'invented' by ancient cyanobacteria some 2.7 billion years ago. Following the logic of the CBC model, this novel sensitivity to light levels was instantiated in cells that were sentient and had a rich sensory system in place. The logical entailment is that all three phenomena (the fundamental sentience of the CBC model, the first circadian clocks, and membrane-based electron transfers) are inherently interlinked. Biomolecular details are in Baluška and

Reber (2021b), Edgar et al. (2012), Loudon (2012), and the issues are explored further in Chapters 5 and 6.

Fatigue

Do bacteria sleep? This is a reasonable question to pose, particularly in light of the existence of circadian clocks. One of the elements of life is the role of down time, taking a break allowing the organism time to recover, re-establish normal metabolic functions, and repair cellular damage. The best evidence to date suggests that the answer is probably not—if we regard sleep as a state with specific neural features marked by shifts in electroencephalograms and with distinct phases like rapid eye movement (REM) sleep that we see in species with nervous systems.

But something like sleep in bacteria may be based on a different biomolecular foundation. As noted, bacteria are sensitive to shifts in ambient light levels and show circadian rhythms. Of the prokaryotes, the most studied are the cyanobacteria. Susan Golden and her colleagues at the University of California, San Diego have shown that the cyanobacterium *Synechococcus elongatus* shifts gene expression rates and physiological processes in response to cycles of light and dark (see Boyd et al., 2016, for details). As noted above, cyanobacteria invented photosynthesis and used it to generate energy. They are also the species presumed to have been involved in the chimerical event that occurred some 1.7 billion years ago when the first precursors of the plant kingdom evolved. We will have more to say about this singular occurrence in Chapter 11 when we look more closely at sentience in plants. Whether these cyclical modifications in biomolecular functions is anything like 'sleep' and whether *S. elongatus* is in a more quiescent state during certain phases of its circadian clock are still unknown. However, there is evidence for quiescence as quasi-sleep at scale. That bacterial quiescent state corresponds to a sharp decrease in metabolic turnover, reproductive rate, and intercommunication.

Summary

We consider these empirical findings to be persuasive and support the original assumption of the CBC: life and sentience are coterminous. All but only living organisms are conscious and have valenced experiences. But, of course, not all agree. As we have noted in more than a few places, the position taken by the CBC is not the standard one in the field—and the situation is not simple. Most researchers in microbiology know the literature and are not only comfortable with our stance but are in general agreement with it—and, in many cases, took similar stances long before we published our work. And, as we noted earlier, many cognitive scientists and neuroscientists are also aware of much of these data, as are many philosophers who have explored the realm of the prokaryotes. But they are reluctant, for a variety of reasons, to acknowledge the core principle—and several of those reasons stand out.

One is based on a model that argues that these behaviours, while exhibiting a remarkable array of functions, are not accompanied by an existential consciousness. They are, so the argument goes, merely robotic processes, driven by genetic factors, and they spin themselves out without awareness, experience, sentience, or a sense of self. There is, in Dennett's phrase, 'competence without comprehension'. We think this argument is wrong, deeply and profoundly wrong. It is not wrong like a messed-up sum, but a wrongness that has deflected the field of what philosopher Ned Block likes to call 'consciousness science' off its proper scientific course. Virtually every recent book on the evolution of minds that we cited in earlier material puts forward one or another version of it. When stripped bare, this position gains nothing in explanatory power and is rife with philosophical and evolutionary biological difficulties.

Let us ask a not-unreasonable question: what could the evolutionary foundation for such a range of adaptive functions be? If each is gene-driven, it means that there would have to have been a stunning array of independently evolved genetic foundations—one for each sensory system, another for interpreting and perceiving what was detected, another for feelings (which, of course, would not really be felt), others for the various forms of learning, additional ones that would establish memories of each acquisitional experience, at least two different ones for communication, yet another for quorum sensing, a separate one for processing temperature gradients, distinguishing different nutrients, directing motility, making decisions about where to relocate, and when to shift gene expressions. It simply makes no sense and, worse, there is no evolutionary mechanism that could create such a stunning array of functions in an organism that is supposedly dumb as a post—especially in the chaotic pre-biotic environment where change, flux, and variation was a staple of the world, and still is. Trying to explain the behaviours of single-celled species as a collection of separate, genetically determined, mindless actions is messy, cumbersome, and unlikely in the extreme. But, toss in the core assumption of the CBC, a palpable sentience, and everything falls neatly into place. We claim William of Occam's crown. We also conclude that 'there is something it is like to be a prokaryote'. We may not know 'what' that is but we are certain that we know 'that' there is.

Second, assuming these kinds of mindless, robotic systems did somehow evolve independent of sentience does not help for the simplest of reasons. It provides no additional explanatory power over the CBC position. And, worse, you are back skewered on the pike of the emergentist's dilemma. If the argument is that sentience was not an inherent feature of even the simplest unicellular species, you have a fundamental responsibility. You need to present a coherent, scientifically testable model that explains how, when, and in what species phenomenal experience, sentience, a mind, first appeared. That, as we have seen, is one tough nut to crack. And, we will repeat yet again: none of the researchers who have taken this position has even acknowledged, let alone attempted, to solve this problem.

Third, taking the CBC stance frees us up from a creeping mysterianism that several scientists and philosophers have recently embraced. Here is an extended quote from a recent book one of us wrote that tried to deal with the issue (Reber, 2019):

> In this world of the new mysterian, the explanatory gap is somehow made real and unbridgeable, the impossibility of matter making mind taken as an epistemic inevitability, accepted as gospel, embraced as truth without evidence. The Hard Problem gets turned into the Impossibly Hard Problem or the Hardest Problem of All.
>
> The Cellular Basis of Consciousness approach saves us from these kinds of mysterian notions. It opens another empirical path to wander down but it's one that has a well-defined goal at the end: uncover the biochemical, molecular processes that first emerged in the simplest life forms that produced the emergence of sentience. After that, it'll just be more evolutionary biology, something we do quite well.

Now it is time to take on the real challenges, the first of which will be to examine the Vitalism–Mechanism debate and look at what we know about the origins of life.

3

What Is Life? The Vitalism–Mechanism Debate and the Origins of Life

What Is Life?

Anyone delving into the definition of life will make a disquieting discovery. There is no established consensus. Enthusiastic scientific arguments are presented on both sides of that contentious debate, and despite our best efforts, there are no unambiguous answers.

Arguably, the most influential contribution to this topic was in 1944 by the Austrian-Irish Nobel Prize-winning physicist Erwin Schrödinger in his seminal publication *What Is Life?* In that compact volume, Schrödinger placed life in terms that make physicists quiver, declaring it to be a special case, defying the second law of thermodynamics. The second law governs the relationship between physical order and disorder, generally represented within the framework of entropy. In an open system, the total disorder of that system must increase over time as increased entropy. Life is exceptional since it seems to contravene the second law as an ordered state that never reaches equilibrium. Consequently, life is a singular physical state. Schrödinger argued that life was defined by this relative 'negentropy' between the external physical world and living organisms, constituting a state of greater structural ordering, opposing otherwise universal entropic expansion.

In the decades beyond Schrödinger's revelatory insight, considerable progress in cellular biology has steadily accrued with our burgeoning knowledge of the internal complexities of competent cells. Pertinently, there is an increasing realization that entropy is interconvertible with information (Brillouin, 1951). From this equivalency, a few particular characteristics of life come into focus. Life is an ordered (entropic) state where information (as an entropic equivalency) has meaning. Crucially, for this conjunction to occur, cellular consciousness must be present. Further, that essential ordered internal state relies on the membranous physical boundary that circumscribes all living cells. The dependence of life on a competent boundary system can now be readily explained. That membranous boundary is the gateway for the flow of

The Sentient Cell. Arthur S. Reber, František Baluška and William B. Miller Jr., Oxford University Press.
 DOI: 10.1093/oso/9780198873211.003.0003

information into all cognitive cells as a foundational aspect of its information management system, whose particulars will be discussed in Chapter 7.

How Did Life Begin?

If the definition of life is elusive, scientists must surely know how life began. Yet, this, too, remains an enigma, stumping even the foremost experts. In 1981, Francis Crick, the co-discoverer of the DNA helix and a Nobel Prize winner, declared his puzzlement, 'An honest man, armed with all the knowledge available to us now, could only state that in some sense, the origin of life appears at the moment to be almost a miracle, so many are the conditions which would have had to have been satisfied to get it going' (Crick, 1981). Paul Davies, a perceptive Professor of Physics at Arizona State University, USA, offered this assessment in 1999: 'Many investigators feel uneasy stating in public that the origin of life is a mystery, even though behind closed doors they admit they are baffled' (Davies, 1999). It remains true still.

Although the exact thresholds that separate the animate from the inanimate remain uncertain, some progress has been made over the decades, establishing those elements considered essential to its formation and directly relating to how complex life on the planet is constructed. Some aspects of the origin of life have achieved a consensus. Nearly all scientists agree that life began as cells. The eminent scientist and essayist Lewis Thomas summed up this perspective:

> The uniformity of the earth's life, more astonishing than its diversity, is accountable by the high probability that we derived, originally, from some single cell, fertilized in a bolt of lightning as the earth cooled. It is from the progeny of this parent cell that we take our looks; we still share genes around, and the resemblance of the enzymes of grasses to those of whales is a family resemblance. (Thomas, 1978)

Many scientists have postulated a 'primordial soup' where life-granting processes might have been stimulated some 4 billion years ago. A Soviet biochemist, Alexander Oparin, believed that atmospheric oxygen would have prevented the synthesis of the requisite organic compounds, insisting that life needed an anoxic period, as a 'primeval soup' of organic molecules for the development of spontaneous life (Deamer, 2020). A pioneer of crystallography and molecular biology, John Desmond Bernal, theorized that spontaneous life required three stages: the appearance of biological monomers, the subsequent development of biological polymers, and the final evolution of molecules to cells. These polymers were crucial since they demonstrate the inherent property of self-replication.

Scientific investigations have revealed that specific properties of organic molecules are necessary for life. Organic molecules exhibit chirality, referring to the non-superimposable three-dimensional forms as mirror images of one another and either right- or left-handedness. For unknown reasons, biological systems are homochiral: amino acids are left-handed, and nucleotides and sugars are right-handed. In typical chemical reactions, the mixture is generally fifty-fifty. Some astrobiologists have postulated that the existence of earthly homochirality is evidence of an extraterrestrial origin for biological molecules, since it is so otherwise exotic on Earth. Curiously, examinations of meteor samples have found amino acids like alanine occurring much more frequently in their levo form, and sugars that are much more common in a dextro form.

One of the major problems with all origin of life theories is that autocatalysis (increasing the rate of a chemical reaction by one of its products) is essential to life. In 1993, the physician and theoretical biologist Stuart Kauffman proposed that life arose from autocatalytic chemical networks, as combinations of the amino acid adenosine, an appropriate ester like a pentafluorophenyl ester, and an autocatalytic adenosine triacid ester. This potent mixture stimulated a prototypical, molecular form of natural selection (Kauffman, 1993). Although an attractive theory as it could account for the self-organization and the proto-state required for self-replication, it suffers from some significant deficiencies. For any biological process to proceed at a living rate, enzymes are required, and their origin must also be explained. Secondly, the presumption that life originated through a process of competitive natural selection is ill-supported since life is sustained much more by cooperation rather than competition (Miller, 2016a; Miller et al., 2020a).

How then might life have been triggered to include its unique combination of molecules and processes? There is no shortage of highly intricate theories. Since this issue remains as clouded today as ever, it is only natural that a welter of competing hypotheses have blossomed to fill the knowledge vacuum:

1. Electric sparks might have generated just the correct amino acids or sugars. Volcanic clouds in the early atmosphere might have held methane, ammonia, and most amino acids that are the basic chemicals of life, and this 'stew' could have led to the self-replication essential for life on Earth (Ignatov & Mosin, 2014). This path satisfies the need for self-replication but says little about the other requisite of self-organization.

2. Pools of condensed and cooled thermal vapours might have started life. A 2012 study at the University of Osnabruck in Germany claimed that adding a type of clay to a mixture of fatty acids yields a form of membranous vesicle formation that could lead to the spontaneous self-assembly of nucleotides, forming RNA on an early anoxic Earth (Mulkidjanian et al., 2012).
3. Scottish organic chemist Alexander Graham Cairns-Smith offered allied 'community clay' theory, indicating that clay minerals could have constituted an appropriate scaffolding for life and catalysed RNA formation (Cairns-Smith, 1971).
4. Others have championed a 'chilly start' hypothesis (Miyakawa et al., 2002). Ice might have covered nearly all of the planet 4 billion years ago since the sun was less luminous than today. Life might have been able to form under this extremely thick layer of ice because the required molecules would have been protected from harmful ultraviolet radiation, giving them more time to stabilize and potentially self-replicate.
5. At the opposite extreme, the 'deep-hot biosphere' model, first proposed by Thomas Gold 30 years ago, suggested that life did not begin on Earth's surface (Colman et al., 2017). Instead, life formed below the surface where even today, microbes with exotic chemistries thrive, with some even using radioactive substances as their energy source.
6. Planetary thermocycling might have done the trick. Life needs an energy source and planetary thermocycling might have triggered thermosynthesis, beginning a process akin to fermentation (Muller, 2005). Protocells might have formed in an energetic convection from a volcanic hot spring, aggregating molecules, stimulating self-organization, and encouraging the heat dissipation, a critical attribute of life.
7. Some origin-of-life scientists insist that 'life's a beach'. The 'radioactive beach' hypothesis postulates that life began in tidal pools where energy-producing radioactive elements might have aggregated and become the energy source for living catalysis (Bywater & Conde-Frieboes, 2005).
8. Possibly, life is all about entropy. Karo Michaelian at the National Autonomous University in Mexico proposes that any robust model of life must account for it as an irreversible thermodynamic process (Michaelian, 2022). Living organisms dissipate energy, equating to entropic production. Conceivably, that entropy production should not be viewed as incidental to life but as its exact purpose and reason for existence. Accordingly, DNA and RNA were formed for the specific purpose of energy dissipation to serve basic thermodynamic processes.

Although this theory might seem attractive to physicists, it subtracts the life out of life for biologists.

9. A very popular theory champions life's origin in deep submarine hydrothermal vents where copious hydrogen-rich molecules spew at or near the sea floor (Martin et al., 2008). Theoretically, congregations of these molecules in rocky nooks might have provided catalytic minerals for critical molecular reactions.
10. For many scientists, biology is all about genes. Naturally, many assume that life arose first from genetic material, either forming spontaneously here on Earth or arriving via panspermia (Line, 2007).
11. Other scientists disagree, insisting that metabolism had to come first. In the 1980s, Günter Wächtershäuser and Karl Popper proposed an 'iron-sulphur world–metabolism-first' model (Wächtershäuser, 1988). Energy for the origin of life derived from sulphides of iron and other minerals such as pyrites, released by reduction and oxidation reactions of metal sulphides to yield organic molecules, including polymers. These molecules became the basis for an autocatalytic chain. Theoretically, this reaction set provided the environment in which RNA replication could emerge.
12. The 'zinc world' hypothesis extends the iron-sulphur and hydrothermal vent models (Mulkidjanian & Galperin, 2009). Fluids with high levels of hydrogen sulphide could have flowed from hydrothermal vents and reacted with cold water to precipitate zinc sulphide, which has the unique ability to store radiation energy such as ultraviolet light. This reaction energized the initial stages of photosynthesis, leading to informational and metabolic molecules.
13. The 'RNA world' hypothesis recognizes that DNA needs proteins to form, but the formation of proteins requires DNA. How could one exist without the other? The answer might have been RNA which stores information just as DNA does, serves as an enzymatic catalyst (ribozyme) as an intermediary in the expression of genetic information, and was likely present on early Earth (Alberts et al., 2002). Although an attractive alternative, one major problem looms. How did RNA form?
14. The 'lipid world' or graded autocatalysis replication domain (GARD) model has no initial requirement for DNA or RNA to jump-start life. In this scenario, the first self-replicating entity was lipid-like (Lancet et al., 2018). For example, phospholipids form lipid bilayers if agitated in water and these micelles are very similar to simple membranes. Further, lipid molecules exhibit hysteresis, a physical property where

molecular conformations lag behind the changes affecting them. This effect can be considered a primitive type of information storage and memory. In theory, these could have evolved into bioactive polymers and initiated a self-replicating cycle.

15. As might be expected, a mixture of many of these have been combined in to a synthesis, now dubbed the 'multiple genesis' hypothesis. The justification is that early Earth chemistry might have been substantially different than it is now and would have gone through many stages, each of which might have had a developmental impact on the emergence of life. Some of these crucial, early chemical compounds might have been subsequently expunged and no longer exist. For example, early life might have been based on arsenic instead of phosphorous. Modern extremophile microbes and their exotic chemistries offer a potential template for these differences. Indeed, it may not have been necessary for any highly complex molecules such as RNA to be present to produce life. In the 'simply beginnings' scenario, life began from small, simple molecules, each interacting with others. Alternatively, there might have been different nucleic acids preceding preceded RNA, the so-called 'pre-RNA world' scenario.
16. Some researchers believe that the origin of life depended on viruses, as a 'virus world' hypothesis (Kostyrka, 2016). Although viruses are presumed to have emerged from cells since they require cellular machinery to replicate, the discovery of giant viruses, such as *Mimivirus*, with more than 1000 genes, has stirred intense debate (Claverie et al., 2006). This virus is so large that amoeba will mistake it for its typical bacterial meal. Pertinently, these giant viruses contain genes for their own replication, can manufacture a few proteins, and can even repair DNA. *Pandoravirus* is even larger and has over 2500 genes, an astounding number when considering that our human complement is approximately 22,000. If true, we not only share a LUCA cell but an additional common denominator from a 'deep viral' past.

Evolutionary biologist Eugene Koonin (2007) focuses on the intractable obstacle within the RNA world hypothesis, which applies analogously to all other theories. Despite many experimental efforts, there is no actual evidence of how replication and translation began. Furthermore, and critically, there is no knowledge of where RNA replicase came from, which is necessary for the translation system that continues beyond it. So, the origin of the core biological processes remains unanswered. Even if there was an RNA beginning, the origin of the essential catalytic enzymes that life requires remains a mystery.

The Vitalism–Mechanistic Debate

Since the beginning of humankind, philosophers and scientists have sought solutions to the riddle of life's origin, either by direct observation or through metaphysical approaches. The ancients believed that life came from non-life since it corresponded so closely with what they could readily observe. For them, abiogenesis was obvious (Sheldon, 2005). To the unaided eye, maggots simply appear on rotting meat, representing the spontaneous generation of life from non-life. Indeed, this was obvious to Aristotle. After all, flies could be directly observed arising from putrid matter, and undoubtedly, frogs emerged *de novo* from slime. The ancient Greeks agreed about the validity of abiogenesis and accounted for the varieties of living forms through 'heterogenesis'. For example, bees came from flowers, not putrid waters.

The belief in spontaneous generation was so ingrained that it persisted for over 1500 years. Even Leeuwenhoek's first microscopic observations of microbes in 1665 was initially interpreted as strengthening that concept since it was believed that microorganisms could not reproduce. It was not until the inspired experiments of Louis Pasteur that the concept of spontaneous generation was replaced by biogenesis—life must issue from prior life. Pasteur recognized that his germ theory would never be fully accepted as long as the belief in spontaneous generation persisted. He devised a simple experiment, putting beef broth in a long-necked flask, and boiling it to sterilization (Gillen & Sherwin, 2008). If the sterilized broth was never exposed to air, it remained unchanged and lifeless. If exposed to air, it became turbid, indicating microbial contamination.

Although Pasteur settled that debate, he raised others. How might first life have arisen? One possibility source gained broad initial credence as a theory of panspermia; life originated elsewhere beyond the planet, coming from another cosmic location. Darwin disagreed and favoured another interpretation, famously opining in an 1871 letter to Joseph Dalton Hooker. According to Darwin, life may have begun in a 'warm little pond, with all sorts of ammonia and phosphoric salts, lights, heat, electricity, etc. present, so that a protein compound was chemically formed ready to undergo still more complex changes' (Damer, 2016). However, for life to take hold in this manner, two possibilities loom. Is 'life', 'mind' or 'consciousness', resulting as a type of special vital spark (vitalism), as a 'vis essentialis' or 'élan vital' that extends beyond materialism? Or alternatively, can life's essence be reduced to a series of discernible materialistic/energetic steps (Masi, 2022)? In other words, is reality 'mind' or 'consciousness' rather than structural matter?

Indeed, this debate has ancient roots in metaphysical vitalism (Masi, 2022). Plato and Aristotle had supposed the presence of an immanent life force, regarded as the soul by Aristotle. The Greek surgeon, physician, and philosopher Galen (129–210) believed in 'pneuma' as an essential life spirit, a concept that continued through the Middle Ages (Masi, 2022). The French biologist and evolutionist Jean-Baptiste Lamarck insisted on the existence of a life-power that conjoined with an inner adaptive force, marrying well with his theory of adaptive evolution (De Klerk, 1979). For many scientists, such as the eminent French physiologist Claude Bernard (1813–1878), it seemed impossible to reduce life to a sum of its parts (Masi, 2022). Life was uniquely intricate, constituting an entirely integrated whole, defying reduction. However, some insisted that this essence need not be metaphysical and might be perhaps conceived as a vital physical process, such as Bernard's homeostasis.

These latter ideas carried substantial weight within the scientific community. Eventually, they gave rise to a scientific neovitalist movement, initially spearheaded by German embryologist Hans Driesch (1867–1941). This form of vitalism was believed by many eminent scientists of that era and earlier, including Pasteur (Miller, 2013). Based on observations of the development of sea urchin embryos, Driesch proposed a reinvigoration of Aristotle's entelechy as an inner vital animating and self-organizing living force (Masi, 2022). This movement enjoyed fleeting popularity but was quickly eclipsed with the rediscovery of the long-forgotten records of Mendel's pea plant genetic experiments by botanists Hugo de Vries and Carl Correns in 1900. From that time forward, vitalism has been discounted as a relic of a former age, tainted by quasi-religious metaphysical convictions rather than based on scientific evidence (Bowler, 2003).

Since 1930, a mechanistic view of life as a living machine has dominated the life sciences. This duelling mind–body dualism stretches from Descartes' 17th-century declaration of animals as 'automata', that he formalized into his mechanical philosophy to explain life (Gorham, 1994). Indeed, the current academic sentiment about the lack of merit of vitalism and derivatively, the categorical rejection notion of any purposeful direction to biological and evolutionary development, has reached such a pejorative that any belief in living purpose or a non-mechanistic approach to living processes is equated with superstition and immature reasoning. In 2010, Finnish psychology researchers studied vitalistic causality in children and adults (Lindeman & Saher, 2007). They concluded that attributing purpose to objects or explaining biological processes in terms of intentionality that might be centred within constructive energetic forces was a common characteristic shared by naive children and superstitious adults. For them, vitalistic thinking was an obvious example of ontological confusion and unmitigated superstition, equating with belief in the supernatural and childish fantasies.

However, there are some scientists who rigorously insist otherwise. Living organisms are not mechanical machines. Theoretical physicist Stuart Kauffman maintains that planetary biology has a set of identifiable living principles that are not in strictly mechanical conformity with known laws of physics (Kauffman, 2000). Others agree that living self-organization and its apparent purposeful agency lies outside the boundaries of our current understanding of the physical sciences (Szent-György, 1972; Prigogine & Stengers, 1984; Gómez-Márquez, 2020).

In our contemporary era, this debate has achieved a new differentiating plateau that clarifies the separation between living entities and mechanical ones. That difference is that life is information based in ways that do not match how information is processed in computers and machines. Consequently, as physicists Sarah Walker and Paul Davies declare, life exists within parameters where 'new principles and potentially even physical laws are necessary' (Walker & Davies, 2017).

There Are Rules of Life

Although we do not have the precise answer to how to define life or how it arose, some rigorous parameters of living processes can be readily identified that determine how life is comprised. Despite all other uncertainties, two critical living essentials are paramount. First, it is not productive to consider life in terms of a 'noun' as a particular assemblage of matter in a strict order. Instead, life is a 'verb' as a dynamic, ongoing, energetic process (De Loof, 2015; Miller, 2020a). Second, life is coterminous with consciousness, and everything in biology and evolution are defined by that cognitive cellular endowment (Miller, 2016a; Baluška & Reber, 2019; Reber, 2019). Crucially, that conscious life is embodied in the cellular form and has remained so for at least 3.8 billion years. Further, we know that the explicit direction of the living process is not directed towards specific forms but is instead a narrative of seeking cellular solutions to environmental stresses to assure continued cellular equipoise and survival (Miller, 2018; Miller et al., 2020a). Further yet, conscious life is capable of sustaining itself by continuously upholding a set of perpetual rules that govern cellular interactions and grant planetary survival, since evidence indicates that the same cellular rules have been in place from the instantiation of life forward (Miller & Torday, 2018).

Current estimates are that life began at least 3.7 billion years ago and possibly as far back as 4.28 billion years (Papineau et al., 2022). A factor that is often underappreciated is that this first evidence of life relies on competent multicellularity for its support. Single cells provide no opportunity for fossilization unless they have adopted a communal, multicellular form that can produce metabolic products that leave fossil remnants. Accordingly, first-life estimates equate with multicellular life. And crucially, this multicellular life exists through identifiable established rules.

There is no reason to suppose that those foundational rules initiated at the beginning of life have materially changed in the intervening aeons, since all life is directed towards sustaining cellular homeorhetic balance and maintaining the crucial self-referential identity that established first life (Miller, 2019). Homeorhesis is a dynamic system that returns to a prior trajectory, standing apart from homeostasis, which is a system that returns to a particular state. Cells are flux agents that maintain states of preferential flow and are never in equilibrium. Cells sustain their preferential homeorhetic equipoise through the defining cellular properties of chemiosmosis and relative negentropy versus the external environment (Torday & Rehan, 2012). Life's living requirement thereby imposes the need for an external membranous boundary that permits those life-defining properties.

What are those further perpetuating biological properties that assure that planetary life thrives? Foremost among these is cooperation. Life willingly cooperates with other life. The reason is that life is an informational interactome and, as will be explained in a later chapter, the cooperative assessment and sharing of information improve its validity (Miller, 2018). Multicellularity is the biological expression of this imperative, further impelling collaboration, cooperation, co-dependence, and generally mutualizing competition (Miller et al., 2019). Consequently, most cells survive by freely trading resources across scales. Cancer cells do not, which is their pathway to competitive cellular destruction.

Cognition/consciousness/sentience weld these cellular 'rules of the road' into biochemical biological expressions. Consequently, two further crucial attributes of life can be readily identified. First, all cellular life proceeds through cell–cell communication (Torday & Rehan, 2012) since critical environmental information must be shared to assure individual cellular self-protection (Miller et al., 2020b). And second, since survival is dependent on conscious, contingent cellular decisions, life is problem-solving (De Loof, 2015; Miller, 2016a). Of the greatest importance, at the scale of cells and even for ourselves, problems are solved through collaboration. Consequently, life and its origins is much more about symbiosis than neo-Darwinian competition (Chapman & Margulis, 1998; Villarreal & Ryan, 2018).

Within this overarching framework, other cellular prerequisites beyond sentience can be identified. Chief among these are the principles of *complementarity* and *recursion*. Complementarity, as a fundamental principle by which 'apparent' opposites unify, was first formulated at the molecular level by Linus Pauling and Max Delbrück as a intermolecular force, and then extended to the cognitive frame by Howard Pattee (see Miller et al., 2019; Baluška et al., 2022b). Although this unification is also a feature of quantum systems, it equally applies to living cells in which information has both a presenting value and a hidden implicate order, the importance of which we will review in a later chapter. Complementarity underscores our living system since cells actively sustain themselves and cooperatively maintain an entire multi-trillion cellular organism and its critical balance, which might otherwise represent a concurrence of opposites (Kafatos, 2014).

Recursion is crucial to sentient cells, enabling the individual cellular assumption of 'as here, so elsewhere'. Collaborative life depends on this sensibility so that groups of self-referential cells can react to information in self-similar patterns (Miller et al., 2019). This fundamental principle assures that differentiated cells as constituents of individual organs remain attuned to one another, following the essential patterns of reiterative self-similarity, mosaic formulation, and fractal reiterations (Kafatos, 2014). All macro creatures are examples of recursion across all scales as the basis of common cellular understanding and effective cell–cell communication. Indeed, the principle of recursion is unarguable. We reproduce to yield self-similar organisms.

Both complementarity and recursion are elemental biological attributes upholding life-sustaining sentience. Both are served by the additional critical principle of *reciprocation* that underscores all multicellular life, either as holobionts or microbial biofilms. All these co-linked processes are crucial to melding the subjective states of self-referential cells to enable seamless multicellular collaborative organisms that can further reiterate into complex living societies and planetary ecosystems (Baluška & Mancuso, 2021; Frank et al., 2022).

Beyond these specific elemental tenets guiding all planetary life, there is an additional set of rules that govern cellular life so that cells can constructively engineer solutions to environmental stresses (Miller, 2016a; Miller et al., 2019, 2020a). Certain laws are absolutely implicit to all the biological activity. Cells uphold the *law of sentience* as agency, cognition, and communication. Furthermore, there would be no evolutionary timeline without the *law of biological continuity* which represents the drive for reproduction, leading to evolutionary space-time. These all link to the *principle of biological learning and memory*, which is dependent on the *principle of biological ordering* (as living negentropy).

As will be fully explained in a later chapter, all multicellularity is driven by concordant cellular engineering. Consequently, complementarity, recursion, self-similarity, and fractal reiterations as mosaic formulations are forces that impel self-consistent biological forms. In toto, these co-linking rules comprise a dominating *principle of biological attraction* as an innate, spontaneous drive for mutual association (Agnati et al., 2009). This principle is best conceptualized as a cohering biological 'attractive' field that energizes the merger of biological entities into greater degrees of complexity. Accordingly, the *principle of biological attraction* underscores the effective measuring assessment, communication, and deployment of information from environmental cues that can be carried into discrete biological expression. The attraction between male sperm and female oocytes that perpetuate multicellular eukaryotic survival is its further manifestation.

The *law of survival (self-repair and self-preservation)* corresponds with this attractive field, enabling organisms to flexibly confront a wide range of environmental confrontations (Baluška et al., 2022b). Together, these two dominating living principles (attraction and survival) constitute a form of cellular glue that motivates the intimate partnerships among cells of many types with varied genomes to survive as a group, yet energetically maintain individual states of homeorhetic preference.

This cognitive-based energetic confluence supports the life-enabling *stable non-equilibrium principle*, first established by Ervin Bauer (1935) as a defence of cellular homeostasis. According to the non-equilibrium principle, an energetically open system is capable of sustaining stable equilibrium separate from the external environment. To support this principle, the cellular membrane acts as a biological Maxwell's demon, decisively contributing to the intelligent cellular maintenance of preferential homeorhetic cellular flux to consistently contradict the second law to maintain every cell as a relatively negentropic state (Baluška et al., 2022b).

None of these foundational principles would be lasting absent one further and all-encompassing determinant. The physicist Walter Elsasser, renowned for explaining Earth's magnetism, contributed mightily to biology by propounding the *principle of constraints* (Elsasser, 1981). The living reality of all biological states is that they exist as a superposition of possible outcomes before any biological expression. Indeed, this is the essence of conscious contingent decision-making, enabling cellular problem-solving and determining whether cells will expend the necessary energy to engage in cell–cell communication. As Elsasser declared, biology is not founded upon mathematical rules. The choices that living organisms make within the vast reservoir of potential states are acts of creativity. However, notably, these can only exert their influence through reiterating constraints as well as liberties. Effective constraints reiterate throughout living systems, extending from thermodynamic limitations to restrictions on protein configurations to imposed physiological limits. Consequently, biology exerts constraints at every level, reinforced through the principles of recursion and recursive self-similarity.

However, none of these principles could have exercised their mandated influence over biological space-time absent yet for a further crucial cellular principle. In order for life to have successfully perpetuated for nearly 4 billion years, it is necessary for life to conform to the *principle of optimality* (Helman, 1986). This dynamic principle was originally identified by Richard Bellmen to identify the best architectural features for computer program design. The optimal path occurs when the initial conditions and control variables established within the first initiating period and solving the first set of problems is followed within all subsequent decisions as solutions to succeeding sub-problems. All computer architecture and derivatively, living systems, are mandated to remain in conformity with first solutions. Accordingly, the state that resulted from the first solution must always be considered to maintain an optimal subsequent trajectory.

Notably, with respect to cellular life, decisions are choices, and those choices must perpetually adhere to the life-sustaining rules that began at the very beginning of life's trajectory, including any foundational quantum states (Igamberdiev & Shklovskiy-Kordi, 2017). This specific living principle explains why all multicellular organisms without exceptions must undergo an obligatory recapitulation through a unicellular zygotic stage for sexual reproduction (see further discussion of this principle in Chapter 8).

Is There Minimal Life?

We have already established that the origins of life and its exact definition remain a mystery, but can a precise threshold of life be established that might give us a clue about the interface between the animate and inanimate? Unfortunately, even here, the evidence is still incomplete. Confoundingly, some molecules behave in such complex ways that it is possible to entertain that they are living subsets within conscious cells. Those same clouded particulars equally apply to viruses. One such 'in-between' molecule is the vital protein, mTOR, the mammalian target of rapamycin. Rapamycin is a naturally derived compound that blocks fungal cell division, originally discovered in a microbe on Rapa Nui (Easter Island). Mammals produce a competitor molecule and another specific protein, mTOR, which is the target of both rapamycin and that competitor (Lieff, 2020). In its human form, mTOR can accomplish some startling things, exhibiting an astonishing range of effects on crucial cellular processes. By sensing nutrients, energy, and oxidation pathways, mTOR participates in the manufacture of proteins by signalling messenger RNA and ribosomes and is necessary for the proper function of the cytoskeleton of the cell. How might one molecule control such a vast array of processes, assess nutrients, oxygen, and energy, and even contribute to the remodelling of the developing human brain? In some ways, mTOR behaves like a brain itself, integrating many highly complex processes and making a wide range of simultaneous decisions. So, is mTOR independently alive? Does it have 'mind'? That has yet to be conclusively decided.

Aside from single molecules, the plasma membrane and the variety of highly intricate intracellular structures, such as the nuclear envelope, centrosome, and the cytoskeletal microtubules, can all act independently or as part of a distinctly choreographed integrated whole, raising the possibility that each independently possesses a minute nanobrain or 'mind' (Baluška et al., 2021b). Moreover, the extraordinary competence of ribosomal r-protein networks, displaying remarkable architectural and functional resemblances to simple nervous systems, suggests that their role in information processing can also be considered a type of nanobrain (Timsit & Grégoire, 2021).

Viruses demonstrate surprisingly competent behaviours, perform complex intracellular tasks, and actively deceive cells. Viruses can cooperate, exhibit contingent behaviours, communicate, share resources, and even associate together in assemblies that resemble social interactions (Diaz-Munoz et al., 2017). These actions are so convincing that an entirely new field of sociovirology has recently developed. Consequently, are viruses alive? Certainly, they behave as if they are, at least within cells. For now, it remains uncertain. However, it is clear that viruses are sufficiently life-like to co-partner with cells for both collaborative and competitive ends. If not exactly alive, they should be considered a liminal form of life whose living expression is context dependent (Miller et al., 2022).

Biology for the 21st Century

> It would seem that 60 years after Erwin Schrödinger wrote his book 'What is Life?' we should be able to answer the question. However, Nature never ceases to challenge the limits of our imagination.
>
> **M. Y. Galperin (2005, p. 149)**

The explicit gap separating the living from the non-living cannot be bridged solely by genetic expression since genes are tools of cells and are inert outside them (Miller, et al., 2020a). Based on many decades of research, we have learned that life is defined more by its processes than its molecules. Masi (2022) considers life as 'anything capable of metabolizing, eating, excreting, maintaining homeostasis, growing, adapting and responding to the environment, reproducing, and evolving'. For most scientists, the capacity to evolve is regarded as the single most important living attribute. In the search for cosmic life, NASA has elevated the capacity for evolution, proclaiming it the central feature in its search for extraterrestrial life. Their definition derives from Carl Sagan's vision of life as a 'self-sustaining chemical system capable of evolution' (as quoted by Masi, 2022). Current research affirms that this definition is entirely insufficient.

Although, the origin of life is unknown, it is certain that it was coincident with the inception of self-referential consciousness. Necessarily, that conscious life exerted its obliged rules to establish and maintain its living foothold. Through these perpetual cellular rules, we are the inheritors of an immense succession of self-referential cellular self-similar reiterations, which have enabled our form of life, level by level.

Part of the defining problem in the study of life's origins is an affliction specific to our modern era. The complexities of every scientific discipline are so overwhelming that each is relatively segregated from others. Accordingly, obtaining consensus within one scientific field or cobbling a cross-disciplinary synthesis imposes natural academic barriers. For instance, in his rewarding book *What Is life?*, Nobel Prize-winner, geneticist, and biomedical expert Sir Paul Nurse emphasizes our need for a modern cell theory in biology. That theory should concentrate on the endowed properties shared by all living organisms, focusing on their competent linkages that enable living organisms to evolve and remain concurrently separate from and connected to the environment (Nurse, 2020). This perspective views life as a propulsive, living process. In contrast, the Harvard Origin of Life project, which includes the exceptional mathematician Martin Nowak, declared that 'life is an engine propelled by evolution' (Humphries, 2013). For Nowak and colleagues, evolution is governed by precise mathematical principles. Indeed, in this framework, evolution is not just an attribute of the living process but is its primary definition, reiterating through combinations of sequences at successive molecular levels (Nowak, 2006). For them, like Sagan, the key drivers of life are evolution and natural selection, which naturally constitute its best definition.

Our approach here, driven by the CBC model, insists otherwise. Life is not propelled by evolution or natural selection. Instead, life is galvanized by the consciousness that resides in all living things. Sentient entities communicate and problem-solve to meet environmental proscriptions (Miller, 2016a, 2018; Baluška & Reber, 2019). This is the timeless living mission of cells. How is a set of problem-solving solutions to environmental stresses communicated? That is heredity in all its forms. Evolution results from this chain of processes that proceeds through self-referential living choices, not the reverse.

We have learned that the basic cell is a complex enactment of physical properties into a reproductive, entirely 'organized whole' as an entrainment of physics, utilizing basic minerals and organic molecules (Ho, 1994). These building blocks may be common in nature, but to join them into living systems requires lipid membranes, sugars, genes, and calcium—all meeting in a compatible forum in which reciprocating combinations can seamlessly merge. How that happened, and its precise order, is yet unknown. Yet, the fundamental alloying force can now be identified. Self-referential cognition prompts biological and evolutionary development.

We regard that recognizing organisms as an 'organized whole' is essential to understanding biology. Indeed, this perspective has stimulated several generations of biologists and philosophers as a form of materialistic holism, often categorized as 'organicism' and separate from vitalism or mechanistic reductionism (Gilbert & Sarkar, 2000). The central tenet of organicism is that the whole consists of its parts but becomes greater than any simple summation of those components because of emergent interactions. The full functions of those biological parts are context dependent, and attempting to separately reduce a component to discrete functions outside its exact context leads to a fundamental misunderstanding of the whole.

This organic holism is increasingly accepted among biologists since it is amply supported by contemporary research, so much so that it fuels much of modern systems biology and the growing interest in emergent self-organization that underpins contemporary studies of metabolism, physiology, and morphology (Allen, 2005). Further, materialistic holism has become an integral aspect of analysing the successful adaptation of organisms within their respective ecological niches (Emmeche, 2004). From our standpoint, organicism is attractive as it does not entail any vitalistic mysticism and avoids the blind ends of standard reductionism. For ourselves, we lay no claim to adherence to any distinct philosophical movement. On the contrary, our allegiance is non-philosophical and biologically based. Life is cognition, embodied in the perpetual cellular form.

We do not yet know how the interface between life and non-life was crossed. The physicist Walter Elsasser insists the behaviour of organisms cannot be reduced to simple physicochemical reactions (Elsasser, 1987). He repudiated 'vitalism', arguing that the structural complexity of cells is 'transcomputational', with computational complexities extending well beyond any machine metaphor. Indeed, he declared that the separation between the living and the non-living represents a 'no-man's land of irrationality' whose intricacies are frankly unfathomable, mirroring the whole universe.

Whatever the case, we can assert that the explicit point of division between the animate and the inanimate rests discretely between chemical processes on the one hand and communication and problem-solving capacity on the other. There may never have been just one line that was crossed, with life emerging on a sliding scale to achieve full cellular competence. Some insist that we are making arbitrary distinctions about what constitutes life. Keep it simple, they assert. Life should be considered just a reproductive lineage capable of collaborating in metabolism (Dupré & O'Malley, 2009). In that circumstance, the concept of life might be appropriately applied to viruses, plasmids, prions, and intracellular organelles.

What we know with certainty is that life requires biological boundaries since the living process is as much a product of its constraints as its liberties. However, once cognition, communication, and problem-solving became entwined within the cellular compartment, the rest of life can be considered a process of self-similar reiterations. Each recapitulation from the first instance of cellular reproduction forward has centred within fundamental cellular principles, reinforced through the accurate cellular assessment of information, enacted through cell–cell communication among competent cells, and always directed towards finding compatible solutions to meet contemporary environmental stresses. Necessarily then, cells represent an informational interactome (Miller, 2018). All environmental stresses must be internalized so that the three basic cellular domains can perpetually meet planetary realities (Miller & Torday, 2018). How well have self-referential cells accomplished this? They have continuously followed their rules for over 3.7 billion years as the planet's only perpetual survivors.

Does this dynamic view of life from the first conscious cell forward constitute vitalism? Surely not in any metaphysical sense. Nonetheless, cellular consciousness is a verifiable reality. That ineffable quality is our living domain and we can securely declare that cellular cognition/sentience is the basal impulse that propels all subsequent biological and evolutionary development. When correctly viewed within this framework, life becomes a property beyond 'being' constituted by material components. Instead, life is a concordant process of self-referential communication, problem-solving, and collaboration. Vitalism or not, that planetary gift of consciousness is the central nexus of our continuous planetary narrative over the billions of years required to yield us. Thus, even if many answers still elude us, the path towards settling the enduring debate between vitalism, organicism, or strictly mechanistic reduction can now be identified. That resolution resides within the identification of those specific factors that enable the elusive, transcendent quality of living sentience conclusively separating the animate from the inanimate, which is the mission of this volume.

4

Emergence and Evolution of Cells

Emergence of First Cells From Ancient Vesicles and Proto-Cells

Life on Earth has evolved for an estimated 4 billion years. For the first 2 billion years of that evolutionary span all life was unicellular, including at first archaea, bacteria, and then eukaryotic protists. The first multicellular organisms emerged some 2 billion years ago and diverged into three main types: fungi, plants, and animals (Porter, 2020; Mills et al., 2022). Based on paleontological and biological studies, our current understanding implies that the most ancient archaea and bacteria invented all the essential cellular metabolic processes on which life depends, including respiration and photosynthesis. All current life on the planet depends on these innovative, primordial processes (Fournier et al., 2021; Gözen et al., 2022). These initiating primal unicellular prokaryotic organisms lacked a cell nucleus but still maintained a relatively simple and stable metabolic repertoire for a long evolutionary span. Through successive waves of slow, continuous evolutionary novelty, archaea and bacteria became increasingly metabolically complex, authoring almost all of the metabolic processes known presently.

The origin of life remains a mystery. Although not the focus of this volume, there are several speculative proposals based on acellular replicators that might account for life including the so-called RNA world hypothesis. It is also possible that life emerged with help of organic compounds brought to the Earth through impacts of asteroids and meteorites known to contain diverse organic molecules that could *seed* life on the Earth (Milshteyn et al., 2019; Oba et al., 2022). Currently, there seems to be overall agreement that the LUCA resulted from the pre-LUCA evolution via the first universal cellular ancestor (FUCA). This early cell evolution was accomplished under anoxic conditions of ancient Earth (Porter, 2020; Gözen et al., 2022). The most plausible scenario for planetary life is that spontaneously formed vesicles merged

The Sentient Cell. Arthur S. Reber, František Baluška and William B. Miller Jr., Oxford University Press.
 DOI: 10.1093/oso/9780198873211.003.0004

together and generated the first proto-cells that could become the competent diverse cell populations that are now present (Tang, 2021; Gözen et al., 2022). These proto-cells perhaps relied on non-enzymatic metabolic reactions (Muchowska et al., 2020).

How the first vesicles emerged is debated. One possible scenario is that early micelles as coacervates (small liquid droplets combining two immiscible liquid phases based on macromolecules with opposite charges or the association of hydrophobic proteins) became the first vesicles (Donau et al., 2020; Gözen et al., 2022). It is proposed that the ancient vesicles had an inherent propensity to fuse together (Donau et al., 2020; Sarkar et al., 2020), thereby representing the prebiotic system prone to develop into the first proto-cells (Kompanichenko, 2012; Spitzer, 2017), perhaps with the assistance of self-replicated RNA molecules (Joyce & Szostak, 2018). Important in this respect is the suitability of ancient vesicles and proto-cells to reach life-specific far-from-equilibrium states by handling both energy and information fluxes via their limiting membranes, allowing the contravention of the second law of thermodynamics that permits the ordered cellular state (Kompanichenko, 2012). This thermodynamic separation of the intracellular space from the extracellular environment is the most critical issue for the initiation and continuation of living processes (Morowitz et al., 1988). Ever since the first excitable plasma membrane was formed, it never again formed *de novo* but remained continuous via replication and has been structurally continuous across 4 billion years of cellular evolution (Fields & Levin, 2018; Baluška et al., 2022b). Relevant in this respect are recent reports on coacervate-like synthetic proto-cells (Abbas et al., 2021; van Stevendaal et al., 2021) which can be spontaneously membranized by amphiphilic silk polymers (Li et al., 2019; Yin et al., 2022) and induced into communications and predatory-like protocell behaviours (Qiao et al., 2017; Grimes et al., 2021). We will further discuss the critical role of the ancient plasma membrane with its dense population of protein macromolecules that are required to handle sensory information and energy fluxes in Chapters 5 and 6.

Cells Are Based on Biomembranes Charged With Excited Electrons and Protons

There is a huge diversity in metabolic processes and sensory signalling cascades which can be illuminated only by bottom-up molecular biology and biochemistry based on experimental analysis. An often-neglected aspect of cellular life is that besides biochemistry and metabolism, crucial bioelectric aspects of cellular life are based on electrostatic phenomena and ionic electrical charge transfers within and between diverse macromolecules (Derr et al., 2020). Saliently, bioelectricity and charge transfers are essential for almost all metabolic processes, especially respiration and photosynthesis invented by ancient prokaryotic organisms and vital for life. This suggests that life evolved via proto-cells leveraging bioelectricity generated by ionic transfers across their limiting membranes. Present-day archaea and bacteria use membrane-based bioelectricity not only for respiration and photosynthesis, but also for cell–cell communication to facilitate stress adaptations (McGlynn et al., 2015; Wegener et al., 2015). Devoted protein complexes push protons across the biological lipid-based membranes associated with electron donor and acceptor proteins arranged in forms of charge-transfer chains, allowing the generation of transmembrane proton gradients charging electrically capable biological membranes (Morowitz et al., 1988; Milshteyn et al., 2019). Bioelectrically charged membranes are then harnessed to fuel ATP synthesis via ATP-synthases spanning the membranes and supporting the active transport of ions (Mitchell, 1961; Kaur et al., 2021). Importantly, biological membrane bioenergetics is universally valid and conserved across all domains of life (Sojo et al., 2014), implying that bioenergetics was the first critical feature of emerging proto-cells allowing the later emergence of the very first cells (Morowitz et al., 1988; Milshteyn et al., 2019). As will be further discussed in Chapters 5 and 6, this specific capacity for cellular bioelectricity and bioenergetics rendered sentient experientiality within ancient cells, facilitating their survival and spurring their further evolution in an unbroken continuum to the present.

Cells use redox-based charge transfers to excite their macromolecules, especially at their biological membranes, allowing the generation of bioelectricity, based on charged transmembrane gradients and voltage-sensitive protein flux across ion channels. Besides excited and movable electrons and protons, diverse reactive molecular species and radicals (derived especially from oxygen, nitrogen, and sulphur) are similarly relevant (Jones & Sies, 2015; Santolini et al., 2019). These free radicals are generated as a by-product of most metabolic processes and actively kept under tight control through complex networks of molecules and processes. Moreover, most of these short-lived radical molecules are an inherent part of homeorhetic signalling systems that maintain stable, preferential cellular states of flux. Homeorhesis represents the tendency for a dynamic system to return to a prior trajectory after being altered by a stimulus. Since all cells are in a constant state of dynamic flux, that term better suits living systems than homeostasis, which implies a stable equilibrium. These cellular flux states particularly pertain to respiration and photosynthesis, involving the release of diverse reactive species by both electron and proton transport processes across biological membranes. Respiration is the much more ancient process and served as a prototype for the later evolution of photosynthesis. That later cardinal event allowed ancient cells to gain independence from Earth-based energy sources and instead, rely solely on the photon-based energy source of the Sun. In that respect, that transition is reminiscent of our current attempts to rid civilization of our dependence on the Earth-based energy sources and oil reserves, making the Sun our major energy source. For that cellular transition billions of years ago, bacteria and archaea used energy-converting hydrogenases for efficient redox-driven complexes as precursors of the contemporary complex I of the mitochondrial respiratory chain (Mühlbauer et al., 2021; Yu et al., 2021).

CBC theory proposes that the emergence of the very first living cells, enclosed with an ancient version of the plasma membrane, is coterminous with the origin of life and consciousness (Reber, 2019; Baluška & Reber, 2019). The cell plasma membrane allowed active and tight control of the intracellular space and generated the essential components of organic chemistry, based on diverse macromolecules, to sustain cellular functions and energy conservation. Survival of these ancient cells and their predecessor proto-cells would only have been possible due to cellular self-referential cognition as cell-based sentience. In order to embark on continuous biological evolution, CBC theory proposes that these proto-cells/progenotes (Woese, 1998; Di Giulio, 2021), had a minimal version of proto-consciousness that was sufficient to respond sensitively and effectively to diverse environmental insults and challenges that were extremely violent on ancient Earth (Knoll et al., 2016; Korenaga, 2021). Cellular proto-consciousness was also required to manage the self-reduplication of their myriad macro-molecules and processes with high levels of complexity during cell division. Unfortunately, no fossil record of these first cells exists. After some 4 billion years, we are left only with the educated guesses and constructive speculations as to the exact capacities of these very first primordial cells. However, one essential component of this critical transition to a fully capable cell is incontrovertible. Any living entity must be thermodynamically separated from the environment in order to generate and maintain internal order that is essential for adaptations, metabolism, and heredity. All of these were necessary to pass the Darwinian threshold of the faithful reproduction of structural integrity and metabolic processes (Woese, 2002, 2004) that is required for continuous biological evolution.

Ancient Bioelectric and Redox Codes Precede Genetic Codes in Cellular Evolution

As mentioned, all life is based on charge transfers by excited electrons and protons. These electrically charged molecules generate dynamic bioelectric fields, most prominently from biological membranes, but also permeating both intracellular and extracellular spaces (Santolini et al., 2019; Derr et al., 2020). These excitable electrons and protons are central to the critical energetic pathways of cellular respiration and photosynthesis that underlie most metabolic processes and mechanisms (Zerfaß et al., 2019). All redox networks were established at the dawn of life with the first, primordial cells which used these pathways to extract energy and information from their environment (Harel et al., 2018; Raanan et al., 2018). Ancient cells still lacked a DNA-based genome but likely used RNA molecules inherited from the hypothetical RNA world for the long-term storage of biological information. The fact that the three basic domains of life, archaea, bacteria, and eukaryotes, have homologous RNA polymerases but different and domain-specific DNA polymerases (Koonin et al., 2020) implies the later acquisition of DNA as a supplemental information storage macromolecule (Di Giulio, 2021). Of greatest importance was the absolute conservation of the crucial bioelectrical and bioenergetic phenomena based on excitable, mobile protons and electrons inherent within the plasma membranes of all archaea and bacteria. Consequently, these phenomena are the ground state for associated cellular sensory perceptions. These are generated first by the plasma membrane receiving environmental informational stimuli which generate code-like patterns of charged particles that move forcefully and meaningfully across the intracellular space stimulating cellular experiences (Adamski, 2016; Levin & Martyniuk, 2018). This specific chain begins with transmembrane membrane flows of signalling molecules and ionic transits that further connect to internal ionic states and integrated cellular receptivity has been productively described as the 'senome' (Baluška & Miller, 2018). The senome represents the totality of the cellular apparatus to assess and measure intracellular information leading to its productive deployment (Baluška & Miller, 2018; Baluška & Reber, 2021a). Accordingly, the formation of a fully competent senomic apparatus is crucial to the emerging sentience of the ancient cells.

Bioelectric and Redox Codes Underlie Cognitive Circadian Clocks and Cellular Sentience

Cellular life emerged on Earth within its predictable rotations constituting our day/night cycles. Ancient proto-cells and the first primordial cells, still lacking a DNA-based genome, evolved their cellular circadian clocks based on their ancient redox code platforms (Cortese-Krott et al., 2017; Santolini et al., 2019). Within contemporary cyanobacteria, it is known that their gene transcription-based circadian clocks are upheld by the ancient redox-based circadian clocks assembled through integration of oxidation–reduction cycles of peroxiredoxin proteins (Edgar et al., 2012; Stangherlin & Reddy, 2013). It can be speculated that emerging vesicles and proto-cells relied on these proto-cellular clocks generated by their ancient excitable, limiting, competent membranes productively channelling both energetic and sensory information-based fluxes (Morowitz, 1978; Wilson & Lin, 1980).

Recently, we have proposed that the cellular evolution of cognitive circadian clocks was closely linked with the evolution of cellular proto-sentience (Baluška & Reber, 2021b). Undoubtedly, ancient proto-cells must have co-evolved their proto-metabolism (Stewart, 2019; Takagi et al., 2020), including their ancient versions of cellular respiration and photosynthesis, to permit the emergence of redox-based sentience and cognitive circadian clocks (Rey & Reddy, 2015; Baluška & Reber, 2021b). Cellular clocks emerged in ancient cells as predictive circuits allowing cognitive cells to productively cooperate and collaborate to further cellular evolution. As we will be discussing extensively in Chapter 6, all electrically charged and redox-based molecules (Cortese-Krott et al., 2017; Santolini et al., 2019) freely move within the available intracellular space, generating dynamic and ever-changing plasma-like and cell-wide bioelectromagnetic fields (Adamski, 2016; Levin & Martyniuk, 2018).

Symbiotic Evolution of Eukaryotic Cells

How the first eukaryotic cells evolved from the ancient prokaryotic cells remains a puzzle but a substantial amount of theoretical analysis has been done. The current consensus is that it took about 2 billion years, a period termed the Lucacene, to evolve the first eukaryotic cells from the conjoining of two previously separate living prokaryotic unicellular organisms (Mikhailovsky &

Gordon, 2021; Nobs et al., 2022). Why might it have taken so long to evolve the first eukaryotic cells? It took just 1 billion years or less to evolve the first prokaryotic cells after the formation of the Earth. From the perspective of the CBC theory, this long span to reach the eukaryotic cell from the prokaryotic cells is strictly related to the imposing task of generating unified eukaryotic cell sentience from several separable prokaryotic cellular sentiences. This complex topic will be discussed in greater detail below and further in Chapter 6. However, before we do so, let us initially focus on the so-called *multicellular* nature of the eukaryotic cell from the perspective of symbiotic unification of originally independent unicellular organisms.

For some curious reason, the symbiotic nature of eukaryotic cells contravened earlier traditional mainstream logic in cellular and evolutionary biology, preferring its entrenched narrative of evolution by gradual adaptive nucleotide point mutations to the self-evident saltationist gaps that evolutionary evidence provides. As a particular example, the concept of a crucial endosymbiosis enabling the eukaryotic cellular form based on mutual symbiosis was greeted by ridicule and scientific dissent when first proposed in Lynn Margulis's 1967 landmark paper, 'On the origin of mitosing cells' (Sagan, 1967). This remarkable paper was rejected by 15 journals, until its ultimate acceptance by the *Journal of Theoretical Biology* (Gray, 2017). The strong entrenched preference for autogenic evolutionary theories had also resulted in the predominant belief that the eukaryotic nucleus was the product of slow cellular evolution through gradual remodelling of intracellular membranes and compartments. In order to keep this autogenic scenario relevant and valid, some inconvenient aspects of the eukaryotic cell biology were conveniently ignored (Baluška & Lyons, 2021). For example, the presence of the centrin-based contractile rhizoplast connecting the basal bodies of microtubule-based eukaryotic flagella with the surface of eukaryotic nucleus suggests that nucleus and flagella have the same evolutionary origin (Baluška & Lyons, 2021; Joukov & De Nicolo, 2019). This mysterious structure is present in numerous sperm cells and many protists including the green algae *Chlamydomonas*. As a heritage of this structure, centrin is present in almost all eukaryotic cells and organisms, having diverse functions related to calcium-induced contractility. The best explanation for the centrin-based contractile rhizoplast (Joukov & De Nicolo, 2019) is an endosymbiotic origin, just as is the case for the origin of the eukaryotic nucleus (Baluška & Lyons, 2021). Endosymbiosis would explain several unique cellular features, including similarities between nuclear pores and cell–cell channels (Baluška et al., 2004; Baluška, 2009), which are also incompatible with autogenic theories of eukaryotic cell evolution.

Multicellular and Archaeal Nature of Eukaryotic Cells

Evolution of the eukaryotic cell still remains mysterious but it represented a momentous event in cell evolution. All present-day eukaryotic cells, including all protists, fungi, plants, and animals, trace back to the last eukaryotic common ancestor (LECA) equipped with a previously evolved microtubular cytoskeleton-based flagellum (Margulis et al., 2006; Koumandou et al., 2013). That early competent eukaryotic cell could engage in an ancient version of phagocytosis for nutrition, resulting in an evolutionary explosion within the eukaryotic cellular domain of life (Koumandou et al., 2013; O'Malley et al., 2019). As already mentioned, the Lucacene transition phase from the LUCA to LECA took approximately 2 billion years which is, obviously, an immense period of time (Mikhailovsky & Gordon, 2021). That extended duration relates to a vital hurdle that had to be overcome. For success, it was necessary to set the internal cellular stage for the accommodation of two different organisms to join together in endosymbiosis between separate previously competent independent ones. To effect that merger, all cellular structures, including their membranes, RNA and DNA polymers, and ribosome-based translational apparatus must have been made compatible with one another for seamless communication and for the sharing of information in order to successfully confront environmental stresses. Most importantly, the immense task of generating a new composite multicellular self-identity as a unique and novel dual agency had to be devised.

In our recently proposed model of emergence of the LECA, two fundamentally different archaeal cells merged, with one becoming the combined cell's nucleus and microtubular apparatus that was the grafted to the actin-based cytoskeleton of the resident archaeal host cell. The result was the 'birth' of the contemporary eukaryotic cells with their rich complement of tools in a combined cytoplasmic space enclosed by one plasma membrane. While the identity of both these unicellular organisms may never be revealed, there are strong recent indications that the host archaeal cell was related to the currently discovered Asgard archaea (Zaremba-Niedzwiedzka et al., 2017; Rodrigues-Oliveira et al., 2023). In contrast, the identity of the putative microtubular guest cell remains unknown. The archaeal nature of both merging host and guest cells would provide some explanation to puzzling issues. For example, it might account for why archaea, in contrast to bacteria, are not pathogenic to eukaryotes (Borrel et al., 2020; Jung et al., 2020) although human and plant archaeomes are abundant (Mahnert et al., 2018; Jung et al., 2020). Furthermore, that association might have provided the conforming channel that permits archaeal constituents as productive aspects of the holobionic microbiomes of both plants and animals.

Slow Emergence of Eukaryotic Cells Due to the Cellular Sentience Binding Problem

The 2-billion-year-long Lucacene period required for the emergence of the first eukaryotic cells (Mikhailovsky & Gordon, 2021) had previously been very difficult to explain. From the CBC theory perspective, however, this lengthy period is logical if cellular cognition and the linked, requisite sentient faculties of the two archaeal cells were based on different structural cell motility platforms. In fact, the host cell cytoarchitecture and motility was based on the actin cytoskeleton whereas the putative guest cell's conscious sense-awareness and problem-solving apparatus relied on the microtubuli-based flagella. As will be further discussed in detail, these two cytoskeletal assemblies are relevant for cellular sentience and their electrical action potentials are closely related to the emergence of supracellular organismal sentience in multicellular organisms. Here we will only briefly introduce the major differences between these two most ancient cytoskeletal systems of eukaryotic cells. The actin cytoskeleton is inherently associated with the action potentials (Baluška & Mancuso, 2019) via its interactions with the plasma membrane based on pulsatile and excitable contractions with cascading reverberations across the cell to its peripheries (Staddon et al., 2022; Yang et al., 2022). The actin cytoskeleton is essential for the exocytosis, endocytosis, and endocytic vesicle recycling, which are all vital for synaptic cell–cell communication (O'Neil et al., 2021; Wu & Chan, 2022). Microtubules are primarily associated with and organized by the microtubule-organizing centres (MTOCs) which are connected with the basal bodies of flagella and cilia, as well as with the nuclear surfaces in cells lacking these structures (Baluška et al., 1997). It is very relevant that both the actin and microtubuli-based cytoskeletal systems represent targets of anaesthetics along with their excitable membranes (Platholi et al., 2014; Granak et al., 2021). Notably, all organisms with a modified cytoskeleton and membranes become insensate after exposure to anaesthetics (Craddock et al., 2017; Hameroff, 2021). As Christof Koch concluded, consciousness cannot be computed just based on information. Instead, consciousness is imprinted and built into specific and dynamic biological structures (Koch, 2018). According to CBC theory, basal consciousness is generated at the cellular level through the coordination of multiple internal structures and processes including the macromolecular assemblies at and around excitable plasma membranes and cytoskeletal polymers. Since all cells have these essential structures, all cellular beings can be anaesthetized at all scales, starting with prokaryotic organisms, unicellular protists, extending to fungi, and all plants and animals, including humans (Baluška et al., 2016; Kelz & Mashour, 2019).

Consequently, the first eukaryotic *supracellular-like* organism required a melding of two hypothetic cellular organisms with two different cytoarchitectures: one was relying on an ancient version of actin cytoskeleton-based cellular sentience whereas the second was based on a microtubular cytoskeleton. These two fundamentally difference cellular sentiences were, at first, distinct from each other. It can be presumed that this integration took a very long evolutionary time span. Ultimately, unified eukaryotic cellular sentience was established and became the new platform for conscious awareness with high levels of responsiveness linking to greater problem-solving abilities. This higher-complexity cellular sentience proved to be very competent, energizing not only an evolutionary burst of creativity leading to a great diversity of protists (Finlay, 2004; Burki et al., 2021; Gooday et al., 2021) but also establishing the proper ground state for the emergence of fully integrated multicellularity which evolved into fungi, plants, and animals.

All planetary habitats were affected, including the dark ocean and deep seafloor that hosts a major part of the Earth biosphere with its huge diversity of pelagic and benthic protists. This abyssal benthic megafauna is still largely uncharacterized and some of these protists, such as xenophyophores, reach huge sizes, up to 24 cm (Gooday et al., 2021). The evolution of true multicellularity that could lead to holobionts (Niklas & Newman, 2020; Umen, 2014), required tight integration of cellular informational systems so that individual cells could seamlessly amalgamate into vast assemblages of trillions of cells of many types and degrees of specialization (Colizzi et al., 2020). For this threshold to be reached, intracellular organellar synapses were required to integrate eukaryotic cells and their organelles (Baluška & Mancuso, 2014). These essential structures play a central role in the informational and mental integration of the myriad eukaryotic cells and their companion microbiome that must coordinate to be fully dextrous holobionts.

Viruses as Key Players in Cellular Evolution Towards Higher Complexities

Viruses remain mysterious, stimulating an ongoing contentious debate about their living or non-living status. Similarly, it is disputed whether viruses preceded cells as part of a putative ancient virus world or if they are an evolutionary cellular product that somehow escaped from diverse cells. Life as we understand it is not possible without a lipid bilayer-based excitable plasma membrane. Consequently, it is more likely that viruses are not alive. However, viruses do have a competent capsid envelope and exhibit many features that are attributed to living entities when they gain entrance to cells and co-opt their internal machinery. Accordingly, it is possible to consider viruses as becoming 'alive' inside cellular boundaries. In this scenario, all viruses are semi-living macromolecules which can escape their *birthplace* within cells, entering a *dormancy-like* inactive stage outside of them much like bacterial spores. However, for their replication, they must return to cells. In order to infect and replicate within the cells, viruses manipulate all the cellular macromolecules, especially their membranes, cytoskeletal elements, and nuclei. In addition, they induce the formation of synaptic-like adhesion domains for their cell–cell spread (Baluška, 2009; Bayliss & Piquet, 2018). Although they were originally not part of neo-Darwinian evolutionary theory, it is clear that they play central roles in cellular and organismal evolution (Koonin & Wolf, 2012). Viruses act as vectors of genetic materials and genes among diverse organisms (Gilbert & Cordaux, 2017; Irwin et al., 2022). Moreover, diverse viral-derived elements are relevant for the evolutionary origins of agency, self, and immunity in all three domains of life. In this regard, viruses represent a vital intermediary of genetic and information transfer and communication across the three cellular domains of Prokaryota, Archaea, and Eukaryota (Miller & Torday, 2018).

Viruses also might play crucial roles in the evolution of the first eukaryotic common ancestor (FECA) and LECA, contributing meiotic genes, as well as genes involved in coordinating the cellular and nuclear merger during eukaryotic gamete fusion, known to predate the emergence of LECA (Ramesh et al., 2005; Moi et al., 2022). Especially relevant to this process is the viral class II fusogen HAP2/GCS1 (Brukman et al., 2022), essential for the gamete fusions in protists (Pinello & Clark, 2022), plants (Valansi et al., 2017), and several other eukaryotic organisms (Podbilewicz, 2014; Pinello et al., 2017). Recently, it was discovered that archaeal fusexins are homologous with eukaryotic HAP2/GCS1 gamete fusion proteins (Moi et al., 2022). Viruses manipulate cellular structures very effectively. They can induce vesicle formation both in and out of cells, either internalizing within or escaping from target cells. Further, these complex actions implicate viruses in the evolution of eukaryotic endocytosis and exocytosis. Interestingly, molecular machinery required for the budding of viral vesicles out of eukaryotic host cells also impact the final stages of the cytokinetic cell division (Carlton & Martin-Serrano, 2007; Callistri et al., 2022), suggesting that ancient viral infections might have been relevant for the invention of cytokinetic cell division of the eukaryotic cell. Notably too, the emergence of Eukaryota was not just dependent on viruses. Archaea was crucial to the emergence of the endosomal sorting complexes required for transport (ESCRT) complex as part of eukaryotic cytokinesis (Hatano et al., 2022). The ESCRT protein complexes have an ancient archaeal origin and are essential in diverse cellular processes including cellular endomembrane organization as well as for viral intracellular life cycles (Meng & Lever, 2021; Callistri et al., 2022).

The path from primordial proto-cell to competent conscious cellular life was the combination of lengthy periods of internal cellular adjustment interspersed with infrequent leaps of evolutionary novelty. Eukaryotic cells, the types of cells that make up our cellular architecture, are the result of a combination of two prior independent living forms learning that mutualistic symbiotic accommodations are their best means of survival. All the internal architectural complexity of cells are tools of that journey, and honing them required billions of years. That entire narrative was dedicated to dual perpetual aims. Each cell acts to uphold its conscious self-integrity by the active, competent, and continuous assimilation of the external environment and thereby grants their perpetual planetary dominion.

5
The Structural and Bioelectrical Basis of Cells

Motto: Life is nothing but an electron looking for a place to rest.

Albert Szent-Györgyi (1972)

The Plasma Membrane as a Smart and Excitable Anti-Entropic Barrier

As we mentioned in Chapter 3, cells are living organisms as open systems, extracting matter, energy, and information from their environment for their maintenance and survival in confrontation with a perpetually changing environment. They export all unwanted material and energy via metabolic degradation products and heat. The first ancient plasma membrane fully enclosed an internal space allowing the assembly of concentration gradients across a proto-version of a competent cellular plasma membrane (Morowitz et al., 1988; Morowitz & Smith, 2007). Disequlibria across the proto-version of that plasma membrane allowed the harnessing of energy and material fluxes via proto-versions of biological Maxwell's demons, as competent sensors, channels, and transporters (Binder & Danchin, 2011; Boël et al., 2019). This effective interface permitted the generation of proto-excitability through electrochemical ionic gradients and out-of-equilibrium bioelectricity-based metabolism. Cellular sentience emerged together with cellular proto-metabolism as an inherent feature of cellular biology at the very beginning of cellular life, allowing ancient cells to embark on continuous cellular evolution based on cognition and sentience. Notably, the intelligent anti-entropic actions of the biological version of Maxwell's demons densely populating the plasma membrane enabled the control of energy fluxes that harnessed the plasma membrane excitability that underpins cellular cognition (Miller et al., 2020a, 2020b), allowing cellular survival and uninterrupted cellular evolution for more than 4 billion years (Baluška et al., 2022b). Bioenergetics based on charged ions, free electrons, and reactive species moving within the

The Sentient Cell. Arthur S. Reber, František Baluška and William B. Miller Jr., Oxford University Press.
 DOI: 10.1093/oso/9780198873211.003.0005

intracellular space is an integral part of the tightly linked senomic apparatus that underlies cellular sentience.

Cellular Bioelectricity Is Based on Charged Ions and Moving Electrons

Cells are based on bioelectricity because life depends on moving electrons. In fact, all cells maintain life by effecting redox reactions in biological macromolecules through the exchange of electrons between electron-donor and electron-acceptor proteins, typically embedded within biological excitable membranes. In addition to the most investigated and understood reactive oxygen species (ROS), underlying both cellular respiration and photosynthesis, there are other reactive sulphur and nitrogen species which are even more ancient and relevant for the evolution of life (Cortese-Krott et al., 2017; Kolluru et al., 2020). Moveable electrons in iron-sulphur cluster-containing ferredoxins appear to go back to the very origin of life (Lane, 2022; Moran, 2022). As we have already discussed in Chapter 4, all reactive species interact generating the so-called redox code (Santolini et al., 2019) as it emerged together with the first cells. This ancient redox code is associated with a similarly ancient electrome code based on excitable membranes and the cytoskeleton (De Loof, 2016; Fillafer et al., 2021). The cellular redox code assembled in the very first cells, emerging from hypothetical proto-cells. This concurrent emergence would explain why our metabolic cycles closely couple with our circadian cycles, both sharing deep similarities (Amponsah et al., 2021; O'Neill, 2021). Intriguingly, a critical role was played by peroxiredoxins which allowed tight co-evolution of ancient circadian clocks and emerging metabolism in the very first cells (Edgar et al., 2012). Oxidation–reduction cycles of peroxiredoxins constitute a universal molecular player within the circadian rhythms of all three domains of life.

Plasma membrane bioelectricity and its associated highly dynamic actin filaments are closely integrated via endocytic vesicle recycling (for plants, see Baluška & Mancuso, 2019; Baluška & Wan, 2021), generating fast bioelectric signals known as action potentials (Beilby, 2007; Hedrich, 2012). Although plant action potentials have a slightly modified ionic background, the properties of action potentials in animals and plants are very similar (Baluška et al., 2021a; Scherzer et al., 2022). In both animals and plants, action potentials are essential for the movements of vital organs and the organism as a whole, and further, both types of movements are blocked by anaesthetics (Baluška & Yokawa, 2021; Scherzer et al., 2022). We will discuss these issues in Chapter 11.

Integrated Cytoplasmic Matrix: Self-Organized and Crowded Protoplasm

Cells are composed of numerous biological macromolecules with proteins being the most numerous and performing the most important tasks. These flow along metabolic, biochemical, biophysical, and signalling pathways, integrating into very complex networks that assemble in the cytomatrix-like cellular interior (Luby-Phelps, 2000, 2013). In several respects, the actions of proteins resemble cognitive counterparts as their conformations and interactions are tightly coupled with cellular cognitive faculties inherently linked to cellular sentience (Baluška et al., 2021b). An excellent example of this linkage are ribosomes, representing the translational apparatus of cells acting at the interface between their genomes and proteomes (Timsit et al., 2021). The crowded intracellular micro-environments, due to dynamically interconnected macromolecules (Rivas et al., 2004; Foffi et al., 2013), can restrict and disturb cytoplasmic diffusion (Kwapiszewska et al., 2020). The cognitive potentials of proteins, their associated macromolecules, and dynamic cytoskeletal polymers (Baluška et al., 2021b; Timsit & Grégoire, 2021) represent a cell-based molecular interactome that has been inherited as a structural continuum across several billion years of evolutionary history, with every successive cell division propagating the initiating structural and functional features of all the ordered biomolecules that comprise competent cells (Baluška et al., 2022b).

The dynamic self-organization of these intracellular structures, organelles, and micro-environments was experimentally documented via vigorous ultracentrifugation of unicellular organisms such as *Euglena*. These hardy microbes remained viable after intracellular stratification of their structures and organelles, recovering quickly from that stress (Srere, 2000). These robust survival characteristics suggest that their highly integrated biological macro-molecules can effectively and rapidly reconstruct ordered networks after mechanical disturbances. There is a natural limit, however, since that self-organizational order is lost in both dying or dead cells (Srere, 2000). Therefore, there must be a highly effective ordering and organizing principle at work inside cells. Our position is that the previously discussed redox and bioelectric codes are involved. We will discuss these important issues more deeply in Chapter 6.

The classical biochemical view of the cytoplasm as *a bag of enzymes* has been proved to be incorrect. Biochemical processes in cells are spatially organized via supra-molecular fractal and hierarchically organized complexes as part of dynamic structural–functional cytoplasmic organization (Aon & Cortassa, 2015; Schmitt & An, 2017). In cellular evolution, ordered macromolecules are inherited through structural templating from lineal ancient cells (Baluška et al., 2023). Spatial assembly of cytoplasmic structures is sensitive to cellular signalling pathways, linking this process directly to cellular sensory-based cognition (Reber & Baluška, 2021; Shapiro, 2021). This inherent and dynamic sensitivity of ordered macromolecules to the environment has been proposed to act as nano-intentionality representing a type of subcellular proto-mind (Fitch, 2008; Timsit & Grégoire, 2021). The concept of nano-intentionality adroitly complements the concept of nano-protoplasts as proposed by Gilbert Ling (2007). We must be aware that the cellular interior of eukaryotic cells is a crowded, active environment based on colloid-like active emulsions and electrically charged hydrogels complemented with diverse biomolecular and cytoskeletal condensates (Charras & Lenz, 2022; Yan et al., 2022). Crowded and integrated composition helps explain the rather surprising survival of naked protoplasts released from wounded coenocytic algae (Kim et al., 2014).

An additional feature of intracellular molecular crowding are biomolecular condensates which are membraneless organelle-like droplets inside the cytoplasm and present in cell organelles, including the nucleus (Li & Jiang, 2022). These self-assemble by liquid-liquid phase separation into transient liquid-like droplets. Importantly, recent studies have revealed that the actin cytoskeleton-based biomolecular condensates orchestrate self-assembly and actively participate in signalling the cell periphery complexes (Charras & Lenz, 2022; Yan et al., 2022).

Actin-Based Cell Periphery, Sensors, Demons, and Membrane-Cytoskeleton Adhesions

The cellular periphery in all eukaryotic cells is supported by the dynamic actin cytoskeleton which is closely connected with sensory signalling initiated through membrane-embedded proteins. These complexes act as specific sensors, representing the biological version of Maxwell's demons (Mizraji, 2021; Baluška et al., 2023), transducing signals vectorially across the plasma membrane towards the cytoskeleton. A special role is played by interstitial water in this connection (Mayer et al., 2006), generating a specific micro-environment that permits all the plasma membrane-based sensors, receptors, ion channels, and other biomolecules that handle biologically relevant information to act as the biological version of Maxwell's demons (Baluška et al., 2023).

Significantly, the specific nature of interstitial water, known as *exclusion zone water*, is sensitive to weak electric and magnetic fields (Rad et al., 2021; Shalatonin & Pollack, 2022) and acts as a target of both local and general anaesthetics (Kundacina et al., 2016). The actin cytoskeleton and microtubules are sensitive to anaesthetics, providing a huge intracellular surface to be associated with the exclusion zone water (Kabir et al., 2003; Chierici et al., 2022). Furthermore, other biomolecules are controlled by their hydration shells in an ultrafast dynamic mode, impacting their structural memories (Elsaesser, 2009; Laage et al., 2017). Bundles of actin filament are sensitive to this molecular crowding and the resulting electrostatic interactions (Castaneda et al., 2021), making the whole actin-based cell cortex prone to the perceptions and amplications of any sensory events effected by the biological version of Maxwell's demons.

There are numerous proteinaceous sensors embedded within the plasma membrane which continuously sense the cellular environment, flexibly changing conformations and passing this sensory information onwards via bioelectric and biochemical signal transduction pathways. During these measurements, cells and their sensors behave as biological versions of Maxwell's demons, acting as a so-called Szilard macromolecule engine (Varn & Crutchfield, 2016). They repeat three-step cycles to extract information from their surroundings through measurement, control, and erasure. Of these three basic stages, the last one is the most energetically demanding. Accordingly, this means that these biological demons are acting in accordance with the second law of thermodynamics. Moreover, the erasure step also leads to resetting the demon for the next measurement within the next cycle. The obtained information is used to support the orderly assembly of macromolecules, representing local *islands* of low entropy submerged within a *cytoplasmic space* of a high entropy (Hoffman, 2016; Varn & Crutchfield, 2016).

Actin-Based Cytoskeleton in Endocytosis and Bioelectricity of the Plasma Membrane

The bioenergetics of the plasma membrane is closely linked to the inherent bioelectrical nature of the actin cytoskeleton. Actin filaments support the electrical propagation of ionic waves (Hunley & Marucho, 2022; Manrique-Bedoya & Marucho, 2022) along large intracellular surfaces of this ancient cytoskeletal system with its deep evolutionary origins in archaea (Akıl et al., 2021). Both actin cytoskeleton and membrane-based bioelectricity are closely linked to the tight control of pH on both sides of the plasma membrane. Moreover, action potentials are associated with cell surface deformations which indicates that bioelectricity is intrinsically linked with both electrostatic and electromechanical phenomena (Galassi & Wilke, 2021). In order to maintain physical and structural plasma membrane homeostasis, eukaryotic cells employ either endocytosis (removal of the excess membrane) or exocytosis (adding membrane from intracellular vesicular stores). The actin cytoskeleton is essential for both processes. Advances in our understanding of the evolutionary origin of action potentials reveal that ancient primordial cells used these fast electrical signals to connect to vesicular trafficking to safeguard the mechanical integrity of these ancient cells (Andrews et al., 2014; Brunet & Arendt, 2016).

Endocytosis is essential not only for safeguarding plasma membrane integrity but also for internalization of the external environment and intercellular interstitial space, generating intracellular *islands* of extracellular space within the cellular interior. This process seems to be unique to eukaryotic cells, contributing to both cellular heterotrophic nutrition and a significant increase in the effectiveness of the sensory border between the inside of cells and their extracellular space. As the limiting membrane of endocytic vesicles and endosomes is derived from the plasma membrane with its dense population of bioactive constituents as the biological version of Maxwell's demons, endocytosis results in cells which are much better informed about their local environment compared to cells with less competent membranes (Baluška et al., 2021b).

In the case of the plasma membrane, an endosomal matrix is assembled from a diverse populations of recycling endosomal vesicles (Sigismund & Scita, 2018), supplementing the redox code in the cellular information matrix, thereby contributing to the vital functions of our recently introduced plasma membrane-based cellular senome concept (Baluška & Miller, 2018). The senome complements the genome and epigenome in the trinity of components that comprise the essential elements of cellular information management systems (Miller et al., 2020a, 2020b). Of these, only the senome is directly linked with the external environment via biological versions of Maxwell's demons embedded within the plasma membrane and plasma membrane-derived membranes of endosomes. Therefore, the senome is directly relevant for cellular cognition and sentience (Baluška et al., 2021b). Sensory signals initiated at the plasma membrane permeate the whole cytoplasmic space and reach the nuclear surface (Matzke et al., 2019). Accordingly, the entire senome apparatus actively participates in the information management system of each cell upstream of any involvement by the cell's central genome. Importantly, these molecular-based biological Maxwell's demons, especially ion channels and transporters, are sensitive not only to the external environment but also to bioelectrical charges within the plasma membrane (Bezanilla, 2018; Hille, 2022).

Bioelectricity-based membranes have been investigated for a long time in neurons, but this field received an unexpected push from the uncovering of the biochemical mysteries of cell energy processes based on ATP. Previously, there was little interest in bioenergetics compared to the study of biological molecules or genes. While the discovery of DNA was anticipated and celebrated, the unexpected discovery of oxidative phosphorylation-based chemiosmotic theory met with a resistant scientific community treating that breakthrough rather like an unwanted child in the family. Consequently, Peter Mitchell's illuminating discovery of a cohering chemiosmotic theory received initial disdain (Prebble, 2001). Mitchell's revelations went against scientific expectations in the same way that Margulis's endosymbiotic theory had. Each took several years before they were accepted (Mitchell, 1961; Prebble, 2001). Eventually, however, Peter Mitchell was awarded the Nobel Prize in 1978. Discoveries by other scientists supporting this new view of bioenergetics were crucial to that acceptance, including the involvement of membranes, protons and electrons, rotary ATPases/synthases, and respiratory complexes (Yip et al., 2011; Sazanov, 2015).

Tubulin-Based Cytoskeleton and Centrosomes: From Basal Bodies to Nuclear Surfaces

In contrast to the actin cytoskeleton, microtubules are widely distributed and intrinsically associated with the basal bodies of flagella/cilia and the nuclear surfaces of multinucleated animal, fungal, and most plant cells. The evolutionary origins of microtubules, centrosomes, and centrioles are tightly linked with the evolution of the eukaryotic nucleus. In ancient protist cells, the nuclear surface is anchored via contractile centrin-based rhizoplasts at the basal bodies of cilia and flagella. Interestingly, sperm cells in animals and lower animals closely resemble hypothetical protist-like cells. These merged with oocyte-like host cells to generate the first eukaryotic cell based on integration of two originally independent cells. Importantly, these two ancient cells were supported by two different cytoskeletal systems: the actin cytoskeleton-based host cell and the microtubules/centrin-based guest cell (Baluška et al., 2004; Baluška & Lyons, 2021).

Endoplasmic Reticulum as a Master Organelle Integrating Other Organelles and Endomembranes

In contrast to the prokaryotic archaea and bacteria, all eukaryotic cells have a complex system of intracellular organelles and endomembranes. There is a clear duality in the endomembrane systems in all known eukaryotic cells—the plasma membrane organizes the outside–inside endocytic membrane flows and the nuclear envelope-associated endoplasmic reticulum organizes the inside–outside endocytic membrane flows. The endoplasmic reticulum acts as a kind of master organelle, making tight membrane–membrane contacts with all other organelles as well as with the plasma membrane (Mathur et al., 2022). These endoplasmic reticulum–organelle contact sites not only integrate the intracellular space of all eukaryotic cells but they also serve as intra-organelle communication platforms (Jain & Zoncu, 2022; Kim et al., 2022) via intracellular synaptic-like membrane adhesion domains (Baluška & Mancuso, 2014).

Reactive Oxygen Species–Calcium Waves Within and Between Cells

The extensive integration of the cellular protoplast is most obvious when signalling events induce calcium and redox waves that rapidly spread throughout cells, often continuing externally to enable supracellular connections through tight synaptic adhesion domains. This feature is well known in neuronal cells, animal tissues, organs (Yuryev et al., 2016), and plant cells and tissues (Tian et al., 2020), as well as within their symbiotic chloroplasts (Frank et al., 2018). ROS represent ancient redox and electrically charged molecules due to their loss and gain of free electrons that tightly couple to crucial cellular metabolic and energetic processes such as the Krebs cycle, respiration photosynthesis, and biological circadian clocks (Pospíšil et al., 2022). All cells actively maintain ROS homeostasis. Part of their signalling and coding roles is associated with intracellular and transcellular ROS waves. In both plants and animals, ROS and calcium waves are tightly integrated with bioelectric action potentials (Lee & Seo, 2022; Scherzer et al., 2022). Redox code and ROS homeostasis are solidly linked to ion homeostasis. Interestingly, neurotransmitter-related molecules, including gamma aminobutyric acid (GABA) and melatonin, play important roles in this respect (Hardeland, 2022). Both GABA and melatonin represent ancient signalling molecules (Xie et al., 2022) and were introduced into evolving eukaryotic cells via bacterial endosymbionts (Reiter et al., 2018; Zhao et al., 2019).

6

The Biophysical Basis of Cellular Sentience

Cellular Sentience Guides Cellular Evolution and Communication

The subject of cellular sentience remains outside of mainstream cell biology despite its strong support by numerous lines of research and observation. One of the major reasons that this topic maintains its relative orphan status is that our scientific analysis of sentience started from the wrong end. Most considerations of sentience begin with the most complex cognitive systems such as the human brain. In science, one should start with the most simple system and, only after some understanding of it at that level, approach more complex systems with greater understanding. There is another significant problem in discussing cellular sentience. Our most important theory in biology, cell theory, is in permanent crisis and confuses the biological connections between cells and organisms.

It is a most important point that although there have been nearly 4 billion years of life on this planet, the first 2 billion years were an exclusive period where the terms cells and organisms were synonymous. There were only unicellular organisms and the first eukaryotic cells emerged symbiotically from several originally independent prokaryotic cells (organisms) about 2 billion years ago. The fact that any single eukaryotic cell is a multicellular assembly representing a *cells within a cell* situation (Baluška et al., 2004a, 2004b) clearly documents the definitional problems facing a unifying framework (Mazzarello, 1999).

An additional serious problem blocking the acceptance of the primacy of cellular sentience is the 'central dogma' of molecular biology formulated after the discovery of DNA. Crick's central dogma states that the information encoded in DNA sequences represents the only source of biological information within cells that is relevant for cellular evolution, development, and morphogenesis. Recently, it has become obvious that transgenerational memory

The Sentient Cell. Arthur S. Reber, František Baluška and William B. Miller Jr., Oxford University Press.
 DOI: 10.1093/oso/9780198873211.003.0006

is possible and represents an often-used pathway, allowing improvements in organismal adaptations to challenging environment (Klosin et al., 2017). This pathway is especially useful in plants which use their sessile lifestyle to permit effective transfer of their sensory stress memories across generations (Byko & Kovalchuk, 2010; Oberkofler et al., 2021). Besides the genome which represents the sequence-based digital storage medium, the plasma membrane and the associated cytoskeleton provide a structure-based templating storage medium as part of the epigenome. A recent study reported that a lipid bilayer membrane composed of phospholipids, but lacking proteins and any other molecules, was capable of long-term potentiation and memory storage (Scott et al., 2022). Cellular sentience supersedes the genome and epigenome, serving as a repository of flexible long-term memory for cells, and is crucial for understanding their proper environmental context.

Sentience is an implicit aspect of cellular cognition and memory and essential for cell survival via contextual decision-making from the very early evolution of cells. Cell and organism were synonymous terms during the first 2 billion years of life. All cells ever existing on this planet, including those of multicellular organisms, have embedded historical memory in their ordered biomolecules. Especially important in this respect, encompassing cellular existence as unicellular organismic forms, is their plasma membranes. This type of membranous historical continuity necessarily became a part of multicellular life, permitting the cells of all multicellular organisms to act as semi-autonomous units still capable of seamless coordination through fast, efficient cell-to-cell communication. This inherent feature of all eukaryotic cells was critical for the communicative emergence of multicellularity, as well as embryogenesis and the development of fungi, plants, animals, and humans. Without an understanding of cells in their historical, organismal, and communicative context, we will not be able to properly understand archaea, bacteria, protists, fungi, plants, and animals. Obviously, without that requisite grasp, our understanding of human diseases and our place in nature will remain compromised.

Plasma Membrane as a Vesicle-Generating Smart Border for Inside–Outside Dichotomy

The plasma membrane permits the establishment and active maintenance of life processes by defining the organismal self (inside) from the extracellular space (outside). This situation is true for all unicellular organisms, but also partially true for all cells of multicellular organisms. The plasma membrane smart border performs a sensory analysis of both the outside space via exteroception and the inside space via interoception. The sensory analysis of the extracellular space via cellular exteroception generates a cellular representation of that extracellular space, whereas interoception allows the generation of a robust internal representation, relevant for the assessment and maintenance of metabolic and senomic cellular homeorhesis. Both exteroception and interoception are essential for sentience-based perception of self whereby cells act as conscious agents capable of problem-solving in their continuous confrontation with their challenging environment in an appropriate contextual manner (Baluška & Levin, 2016; Reber & Baluška, 2021). In order to survive, cells need to understand contemporary environmental challenges through both a historical and current environmental context. Necessarily, cells are historical systems which continuously generate and retrieve cellular long-term memories and knowledge of their complete evolutionary narrative and store them in their epigenomes and genomes to enable cellular problem-solving and prediction (Baluška & Reber, 2019; Baluška et al., 2022b).

Among its many functions, the plasma membrane is a universal vesicle generator system wherein vesicles are pinched-off from the plasma membrane extending into the cellular interior and exterior. These events allow structural and functional homeostatic balance of the plasma membrane, despite a heritage of turbulent evolutionary phases at the coacervate–protocells stages. Internalization of small vesicles is known as endocytosis. These endocytic vesicles can merge together into endosomes, which can themselves pinch-off vesicles that can later fuse back with the plasma membrane. This process is very important for neuronal cell–cell communication and these recycled neuronal vesicles from the plasma membrane are known as synaptic vesicles which can be released into the extracellular space.

Importantly, the limiting membranes of both intracellular and extracellular vesicles are each derived from the plasma membrane, and accordingly, all have relevant features and properties that generate senomic fields. The previously discussed biological versions of the Maxwell's demons that populate the plasma membrane are active in these vesicles, generating mini-senomic vesicular fields that expand the plasma membrane-based senome both intracellularly and extracellularly. Consequently, in cells with active endocytosis, such as neurons and root apex cells, the limiting membranes of endosomal vesicles significantly magnify the reach and scope of the plasma membrane-based cellular senomic system. We speculate that the vesicular senomic fields integrate together with the plasma membrane-based fields to form a robust cell-periphery N-space senomic field which can interact with those of neighbouring cells in multicellular organisms.

Further, these plasma membrane-based senomic fields are complemented in eukaryotic cells by organelle-based subcellular fields and with micro-senome fields emanating from diverse macromolecules and cytoskeletal polymers. Particularly strong senomic fields can be expected to be generated by the nuclear envelope and endoplasmic reticulum of each cell. The limiting membranes of these organelles are associated with other critical intracellular participants, including peroxisomes, mitochondria, and plastids. Most of these organelles release vesicles into the cytoplasmic space which again support the entire cellular senomic matrix, allowing cells to communicate and interact together at a distance which is especially relevant for immunity (Tung et al., 2018; Buzás, 2023) but also plays a role in carcinogenesis (Wortzel et al., 2019; Zho et al., 2021).

From an evolutionary perspective, it is important to emphasize that there are three cellular sources of senomic fields in eukaryotic cells. As we have discussed in Chapter 4, in the endosymbiotic merger that enabled eukaryotic cells, the actin cytoskeleton-based host cell is represented by the plasma membrane and the cytoplasm whereas, the tubulin/centrin cytoskeleton-based guest cell is represented by the nucleus and cilia/flagella. The third cellular source of the eukaryotic senome is provided by their endosymbiotic organelles, descended from previously free-living cellular organisms.

The Plasma Membrane-Based Senome Provides the Biophysical Basis of Cellular Sentience

The plasma membrane is inherently excitable, providing not only the structural but also the bioelectric boundary of the cellular self (Veech et al., 2002; Oliveira-Brett, 2017). This feature of the plasma membrane and cell periphery was discovered, among others, by Ernest Everett Just who studied excitable egg cells (Just, 1939; Byrnes, 2020). As all cells of an organism derive from a fertilized egg cell, so all present cells derive from a continuous chain of cell divisions extending back to the very first cells; they retain semi-independent organismal heritage and agency even within multicellular organisms (Baluška et al., 2022b). While the prokaryotic senome is relatively simple, eukaryotic cells assemble a very complex cellular senome by deploying abundant endomembranes. The most critical issue for the assembly of the senome and cellular sentience is maintaining a complete isolation of the intracellular space from the extracellular environment. As we have discussed earlier, the slow and turbulent evolution of the ancient proto-plasma membrane surrounding coacervate-like proto-cells was associated with the very slow emergence of the cellular membrane- and senome-based sentience, building complexity over time. Accordingly, the most simple senomes are associated with vesicles which are also abundantly present within all eukaryotic cells and are considered to be more recent.

With vesicles generated by endocytosis, there is unique inverted topology since these vesicles enclose what was originally an outside space. This somewhat paradoxical feature is further complicated in endosomes, known as multivesicular bodies. In these complex structures, a primary limiting membrane originates from the cellular plasma membrane which then releases smaller endovesicles into the lumen of each multivesicular body. By fusion with the plasma membrane, the interior vesicles of these complex endosomes (vesicular bodies) are then released into the extracellular space as extracellular vesicles. It can be expected that all these vesicles contribute to the senomic complexity of the eukaryotic cell functioning in many capacities including serving as a type of environmental exploration by the cell. All cells produce extracellular vesicles, including bacteria and archaea (McMillan & Kuehn, 2021; Buzás, 2023). Such vesicles represent the smallest senomic units with a wide range of activities including cell–cell communication and even participating in cancer cell invasions. Intriguingly, such extracellular vesicles might represent vestiges of proto-vesicles that preceded hypothetical proto-cells in the evolution of the first cells. Furthermore, extracellular vesicles play important roles in cancer biology by changing cell and tissue identities (Gopal et al., 2017; Zhou et al., 2021). Future studies on the extracellular vesicles circulating between cells will be essential for our improved understanding of embryogenesis, development, and cancer. Extracellular vesicles serve as a primary means of inter-kingdom communication (Cai et al., 2021; Díaz-Garrido et al., 2021). All cells are known to release extracellular vesicles, including archaea, bacteria, and algae (Liu et al., 2021; Picciotto et al., 2022), thus, these vesicles have a huge impact on maintaining cellular ecologies and the entire course of biological evolution.

Membrane-Based Senomic Fields Permeate Cellular Insides and Radiate Out of Cells

The redox-based homeostasis and excitable bioelectrical capacities of the plasma membrane generate an extracellular version of the cellular senome that projects externally for some distance beyond an individual cell. At the present state of our knowledge, we have limited information about this process. However, measurements of extracellular electric fields reveal that all cells, including archaea, bacteria, and all eukaryotic cells in multicellular organisms, generate bioelectric and biomagnetic fields at their surfaces (Wegener et al., 2015; Fabricant et al., 2021). These cellular senomic fields (Baluška & Miller, 2018) integrate into supracellular senomic fields, acting in collective aggregates—N-space Episenomes (Baluška & Miller, 2018; Miller et al., 2020b). The excitable plasma membrane not only generates these dynamic biofields but is reciprocally sensitive to them (Galassi & Wilke, 2021), enabling memories to be stored (Yang et al., 2020) and communicated (Manna et al., 2020) via dynamic supracellular senomic fields. For instance, in bacteria, these fields are critical to the formation of biofilms. Bio-electromagnetic fields extend relatively far beyond any single cell and can be easily measured online using microelectrodes (Jaffe & Nucitelli, 1977; Portes & Feijó, 2021). These same fields extend out from organelles (Matamala et al., 2021; Klier et al., 2022), and can be further identified around sensoric organs such as leaf traps of carnivorous plants (Fabricant et al., 2021), and growing root apices (Collings et al., 1992; Masi et al., 2009). Furthermore, these fields have documented roles in the cognitive function of animals, including our human brain (Buzsáki et al., 2012; Hales & Ericson, 2022).

These same fields can aggregate as supracellular bio-electromagnetic fields helping to explain such phenomena as the so-called peripersonal space which is meaningful in the social organization of humans and other organisms (Salomon et al., 2017; Serino, 2019). There may also be other bioavailable field sources. For example, besides bio-electromagnetic fields, so-called reactive clouds and oxidation fields might be relevant in explaining many biological phenomena (Zannoni et al., 2022; Schoemaecker & Carslaw, 2022). We will discuss these issues further in later chapters dealing with the N-space Episenome.

Macromolecular, Cytoskeleton-Based, Organellar and Vesicular Senomic Micro-Fields Integrate With the Plasma Membrane-Based Cellular Senome

Every cellular macromolecule and all polymers are exquisitely dynamic and sensitive to any signals entering into cells from the external extracellular space. These also vibrate and oscillate, sending out waves that permeate the entire cellular senome field, functioning as electromagnetic ripples. Proteins are also known to represent dynamic and highly flexible structures based on short-range electron transfers and electrostatic interactions within their structures which are very sensitive to electromagnetic ripples and vortices in the intracellular environment (Sjulstok et al., 2015; Tsai et al., 2015). Moreover, contractile properties of cytoskeletal polymers also generate electromagnetic micro-fields and are also sensitive to cytoplasmic macro-fields.

As the cytoplasm is a very crowded environment (Ellis, 2001; Knapp & Huang, 2018), all these senomic sources dynamically integrate into a unified senomic field whose major source is the excitable plasma membrane densely populated with the biological version of Maxwell's demons including diverse receptors, sensors, and ion channels. Cytoplasmic crowding is an ancient feature, typical in small bacterial cells and actively maintained by the cellular activities within all of the cellular domains (Parry et al., 2013; van den Berg et al., 2017). Interestingly, as will be discussed in Chapter 11, physical pressure relieves effects of all known anaesthetics on all tested organisms, suggesting that cytoplasmic crowding is relevant for the cellular sentience.

As vesicles are enclosed by a plasma membrane-derived smart and flexible border, their inside spaces can be expected to be permeated with radical molecular and ionic charge-based senomic vesicular micro-fields as a further derivative of that main field. Such vesicles contain diverse proteins and RNA molecules and are internalized by their target cells via endocytosis. It would be expected that these vesicular micro-senomes would affect the senomes of adjacent cells and have an impact on collective sentience and action. The potential range of effects of these small senomic vesicular messengers on the generation of integrated sentience of multicellular organisms will be discussed later.

The extracellular vesicle emerges as an important player in innate and adaptive immunity (Buzás, 2023). Almost all of the diverse cells of the human immune system produce extracellular vesicles which serve, among other potential roles, in cell–cell communication in their complex task to discern self from non-self cells and defend humans and other multicellular organisms from pathogen attacks. Immune cells producing diverse extracellular vesicles are also involved in embryogenesis, development, regeneration, and homeostasis of multicellular organisms (Taubner, 2013; Underhill et al., 2016). Ilya Metchnikoff's discovery of the cellular basis of human and animal immunity as well as of phagocytosis accomplished by amoeba-like motile cells was a unique advance in cell biology, evolutionary theory, and medicine (Taubner & Chernyak, 1991; Taubner, 2003). In future, it will be important to explore senomic micro-fields of extracellular vesicles and their roles in immunity as well as in embryogenesis, development, regeneration, and homeostasis of multicellular organisms.

Rotary ATPases as Ancient Bioelectric Macromolecular Devices Charging Membrane-Based Biological Batteries

The bioelectric drive behind the excitable prokaryotic membranes responsible for both cellular respiration and photosynthesis are the multi-protein enzymatic rotary complexes known as ATP synthase (Yoshida et al., 2001; Kühlbrandt & Davies, 2016). Rotary ATPases are ancient nanodevices that couple the translocation of protons across membranes to accomplish ATP synthesis or, as in the case of bacterial F-ATPases, can run in the reverse mode to enable ATP hydrolysis (Nirody et al., 2020). There are several types of ATP synthases: prokaryotic A- and F-type ATP synthase (A- and F-ATPase) present in archaea, bacteria, mitochondria, and plastids (Hahn et al., 2018; Nirody et al., 2020), and eukaryotic vacuolar ATPase (V-ATPase), which energizes endosomal vesicles, endosomes, and vacuoles. The complexity of these ancient macromolecular assemblies capable of electrically charging biological membranes is breathtaking (Kühlbrandt, 2019). These complexes are composed of more than 22 molecular subunits (Nakamoto et al., 2008) arranged in a characteristic way by connecting two rotary motor-like domains (R_0 and R_1) coordinated via a central stalk domain (Stewart et al., 2013; Kühlbrandt, 2019). The membrane embedded R_0 motor domain rotates within the membrane plane very fast, acting as a biological nano-turbine, with one rotation releasing three ATP molecules (Fillingame, 2000). These nano-turbines rotate at astonishingly high speeds of more than 400 rotations per second (Nakanishi-Matsui et al., 2010) and show very near 100% efficiency (Kinoshita et al., 2000; Saita et al., 2015). Surprisingly, these rotary ATPases have a common evolutionary origin with the flagellar export apparatus (Ibuki et al., 2011; Ishikawa et al., 2013).

Besides the membranes of bacteria and archaea (Müller et al., 2005; Grüber et al., 2014), endosomal and vacuolar membranes are energized via structural and evolutionarily similar rotary ATPases known as V-ATPases (Collins & Forgac, 2020; Kühlbrandt, 2019). These have crucial roles in ionic and pH homeostasis of eukaryotic cells with important effects on virtually all aspects of cellular biology, ranging from cellular respiration and photosynthesis up to neuronal signalling, synaptic transmission, and cancer cell induction (Ko et al., 2020; Gao et al., 2022). V-ATPases within endosomal, lysosomal, and synaptic vesicle membranes also play a central role in phagocytosis and symbiosis (Mills, 2020), thereby having direct connections to the evolution of eukaryotic cells (Yutin et al., 2009).

Unique Roles of Interface Water in the Biophysical Basis of Cellular Sentience

Water represents the major part of cells. However, inside cells, water almost never exists as pure water. Instead, the cellular interior consists of complex forms of water due to numerous electrostatic interactions with myriad diverse, and electrically charged biological molecules and macromolecules (Fayer, 2012; Ball, 2017). Albert Szent-Győrgi called water the matrix of life (Ball, 2017) and specific hydrophilic and hydrophobic properties of phospholipids provide the structural and functional basis for cellular membranes. As already mentioned, pure phospholipid-based bilayer membranes can store memories via long-term potentiation phenomena (Scott et al., 2022). These membranes are formed by self-assembly of a phospholipid bilayer following the basic rules of thermodynamics for organizing membrane lipids and proteins with associated cytoskeletal polymers (Nicolson, 2014; Nicolson & Ferreira de Mattos, 2022). Many unique aspects of life ultimately reflect specific features of water which are still far from being fully understood. In particular, water situated close to interfaces with internal macromolecules and polymers, known as interfacial water, has unique biophysical and biochemical properties (Messori, 2019). This water has higher viscosity and ordering than pure water, preventing the diffusion of even small molecules due to its semi-crystalline nature. This interfacial water was discovered in 1978 (Mollenhauer & Morre, 1978) and later investigated, especially by Gerald Pollack's research group (Pollack, 2001, 2013; Kowacz & Pollack, 2021). Intriguingly, this unique state of water is sensitive to weak electric fields induced by magnetic fields (Rad et al., 2021; Shalatonin & Pollack, 2022). An important role in this respect is also played by ATP which is the most ancient and abundant component of cells (Chu et al., 2022). Besides its best known role as the cellular energy currency, ATP is also acting with interfacial water as a biological hydrotrope (Patel et al., 2017; Rice & Rosen, 2017). ATP maintains dynamicity of cellular processes at the nano-level by preventing macromolecular aggregations in the crowded cellular interior (Bye et al., 2021; Sarkar & Mondal, 2021).

As discussed by Peter Hoffmann in his excellent book *Life's Ratchet: How Molecular Machines Extract Order From Chaos* (Hoffmann, 2012), water molecules are electrically charged and electro-negative oxygen molecules bind electro-positive hydrogen molecules through hydrogen bonds, representing one of the molecular entropic forces. Importantly, molecular exclusion zones and molecular assemblies are playing crucial roles in these biomolecular entropic forces (see Figure 4.2 in Hoffmann, 2012). This is one of the most important roles of the intracellular biomolecular crowding. We will discuss these topics in deeper details in Chapter 10 dealing with the cellular targets of anaesthetics.

Electrostatic interactions play a crucial role in dynamic assembly of electrically charged biological membranes and their association with internal cytoskeletal elements and polymers of extracellular matrices (Platre & Jaillais, 2017; Ben-Amotz, 2022). Actin filaments and microtubules are particularly important for intracellular propagation of ionic waves (Hunley & Marucho, 2022; Manrique-Bedoya & Marucho, 2022), which act as some kind of intracellular neural system. All these bioelectric interactions establish the so-called cellular electrome (De Loof, 2016) and metallome (Galera-Laporta et al., 2021), which are part of the senome which acts as a non-genomic system upstream of the cellular genome in the management of biological information. In order to correctly understand organisms and living properties, we must turn back to cells, understand them in their historical context, honour their cognitive faculties, and, in particular, we must put cells into the proper senomic framework, correcting the so-called central dogma of molecular biology (Crick, 1970). Of the utmost importance, and quite contrary to conventional thinking, the primary role in cell biology is played by the cell's intelligent, competent biomembranes and their associated bioelectric fields rather than by their genes.

Cellular Electrome and Sensitivity to Extracellular Electric Fields

Existence of the cellular electrome is strongly supported by the documented generation of bioelectrical fields around cells and cellular organs that is associated with the extreme sensitivity of cellular structures and processes to externally applied electric fields. It is certain that the plasma membrane is the most ancient cellular structure. Further, all biomembranes of currently living organisms are directly and structurally linked to an initial primordial proto-membrane that preceded the DNA-based genome even though most biologists still consider the genome as the most important cellular system underlying all subsequent life. This ingrained attitude is partially linked to the central dogma of molecular biology defined by Francis Crick more than 50 years ago (Crick, 1970). Nonetheless, recent studies clearly document that development, embryogenesis, and regeneration can be controlled by extracellular electric fields and that the endogenous electric fields represent a primary bioactive mechanism regulating organismal morphogenesis (Levin, 2021;

O'Hara-Wright et al., 2022). Thus, the central role of the bioelectric code-based senomic electrome must be explored through future experimental studies, seeking to better comprehend those processes that control cellular life, cellular behaviours, and planetary evolution, ultimately granting material insight into our human living predicament and the diseases that afflict us.

As in the conclusion of Chapters 5 and 6, cellular sentience is not based on mysterious vitalistic forces but rather on molecular biophysics based on classical as well as exotic physics waiting to be explained and understood. As cellular biophysics underlies cellular biochemistry and current molecular biology, it will be essential to study cellular biology from nano-scale physical perspectives. We badly need a fresh new biophysical approach to biology by reconnecting to Max Delbrück's insightful analysis from 1949 (Delbrück, 1949; Strauss, 2017).

7

The Biological Information Cycle

The Terms of Consciousness

> Developmental biology can be seen as the study of how information in the genome is translated into adult structure, and evolutionary biology of how the information came to be there in the first place.
>
> **John Maynard Smith (2000)**

Introduction

In Chapter 3, the problems of defining life and precisely understanding its origin were discussed, and the various factors delineating the living process were reviewed. One intriguing means of exploring how the animate separates from the inanimate resides in the complex topic of information science. Pertinently, there is a clear-cut distinction between how computers and machines use information and how living organisms assess and utilize it. That crux is that all information is ambiguous in the living state and thereby distinctly unlike the zeros and ones that constitute computer data.

This critical difference explains several previously unresolved issues in biological development, including the origin of multicellularity and the role of the virome in evolutionary biology. Furthermore, it places life in its correct perspective. The most important aspect of biology is that all cells are conscious, self-aware, self-referential, sentient agents. Living organisms are not automatons that receive information from the environment and respond to it automatically according to a fixed programme. Instead, intelligent cells measure the information they receive, which is the essence of our biological experience (Miller et al., 2020a). Cells measure information to assess its meaning and, based on those measurements, decide whether or not to communicate that information to other living organisms and deploy its meaningful content as biological expression (Miller, 2018). Since the living state is dependent on that

The Sentient Cell. Arthur S. Reber, František Baluška and William B. Miller Jr., Oxford University Press.
 DOI: 10.1093/oso/9780198873211.003.0007

cellular measurement of information, it is apparent that life represents an informational interactome and that the maintenance of cellular integrity and survival is rigorously dependent on the competency of its information management system. This chapter will explore this intimately linked cellular chain of events and explain its further implications.

How Is Information Defined?

At first consideration, most would expect that defining information would be straightforward. Everybody is familiar with the term. We commonly refer to it every day and understand what we mean by its meaning in casual conversation. However, a certain level of rigour is demanded when science deals with any issue. In scientific terms, that stipulates a search for a consistent meaning of information that can be methodically measured and compared. There is no problem with that requirement in general terms. However, as is commonly affirmed, 'the devil is in the details'.

As was previously discussed relating to the definition of life, it is surprising that a general consensus about what constitutes information has proved similarly elusive. The term 'information' can be traced to ancient and medieval texts, but it was only in the 20th century that any attempts at a formalized definition were launched, particularly through the work of Claude Shannon (1948) and the inauguration of the concept of Shannon information. Surprisingly, the Shannon information concept is not a specific theory of information. Instead, it is a theory of communication (Schroeder, 2017). Shannon's formalism provided a mathematical definition of information designed to solve problems in communication and engineering, and is particularly suited to computer systems and the search for optimal coding and robust communication. Shannon information (entropy) relates to the statistical properties within a given system and how states between different systems correlate. This definitional stance is fine for engineering applications but has no direct bearing on the qualitative aspects of information that we depend on in our everyday discourse, remaining devoid of semantic intent and independent of its qualitative value (Schroeder, 2017). Consequently, computer scientists and communications engineers view information in statistical terms and mathematical parameters, and utilize formal definitions that are highly suitable to specific applications, such as Fischer information (random variables influenced by unknown parameters), algorithmic information used in computational programming, or quantum information as a statistical measure of von Neumann entropy that has direct bearing on calculating quantum resources (Lombardi et al., 2015).

By design, these approaches to information are closely tailored to discrete tasks and thus do not match the requirements of a more broadly conceived definition of information that could better suit the variabilities of biological systems. Given the sophistication of each of the disciplines within physics and engineering, the specific definition of information that is employed is dependent on its application within that discipline. Hence, it is not unexpected that there is no general cross-disciplinary consensus as to a precise meaning for information in biology. For example, in computer applications, the term 'information' translates into 'data' and the language of bits and bytes, where a 'bit' is generally conceived as the minimum data unit that describes a 'single difference' in that application (Farnsworth et al., 2013). Although these definitions are superbly helpful for designing robots, they do not meet the need of understanding the nuances of information in living systems.

To better conceptualize living systems, other scientists have undertaken different approaches. For example, the British physicist David MacKay stated, 'Every piece of information has the characteristic that it makes a positive assertion and at the same time makes a denial of the opposite of that assertion' (Schroeder, 2017, p. 2). Of course, he was a renowned mathematician, and what seems quite plain to mathematicians and physicists can be difficult for others, like ourselves, to grasp. However, he was trying to place information into a formal logic of 'representation' that encompasses patterns, pictures, or models, specifically as concrete forms or abstractions that symbolize an element of meaningful content corresponding to the structure being considered. Central to this perspective is the concept of 'difference' that can be directly applied to distinguishing differences between objects of thought.

The polymath and anthropologist Gregory Bateson is generally credited for offering the broad definition of information that is used most often to describe biological activities. In his 1972 book *Steps to an Ecology of Mind*, Bateson developed a synthetic definition of information based on his work on semiotic messaging and cybernetics, stating that information is 'a difference that makes a difference' to which he later amended the codicil, 'in some later event' (Schroeder, 2017). For Bateson, these differences were crucial to formulating a 'map' of the abstract information that the mind constructs. Adding to the work of MacKay and Bateson, the American philosopher Frederick Dretske offered a flexible definition of information that is particularly relevant to living systems, emphasizing that 'information is a commodity that, given the right recipient, is capable of yielding knowledge' (Dretske, 1981).

Biological Information Is Different From Computer Data

In computer systems, data can be readily moved and transferred as binary functions, characterized by sequences of 'ones' and 'zeros'. However, for computer data to become information, it must be interpreted, following a pre-programmed set of instructions. Significantly, there is no assessment of computer data for qualitative aspects since, by design, it is meant to be precise and uniform so that computer outputs will be exactly reproducible for each set of identical inputs.

However, this form of information content is vastly different from information in the living state. For living entities, all information is ambiguous (Miller, 2016a, 2018; Today & Miller, 2017). Living information has both quantitative and qualitative aspects, unlike computer data. The differential crux is that living organisms are cognitive entities and computers are inanimate machine mechanisms. Machines use digital, binary data and living organisms primarily use analogue information, although some geneticists regard the genome as a quasi-digital information system. One way to visualize the critical differences between these two forms of information is to examine simple waveforms that differentiate digital data from analogue information (Figure 7.1).

A digital waveform is characterized by a square wave, denoting its binary character, reflecting discrete orthogonal signals mirroring how digital electronic circuits instantaneously respond. An analogue waveform is alternatively depicted as a smooth continuous sinusoidal wave that depicts harmonic averaging. Consequently, analogue systems are much more susceptible to noise and are more likely to produce significant errors.

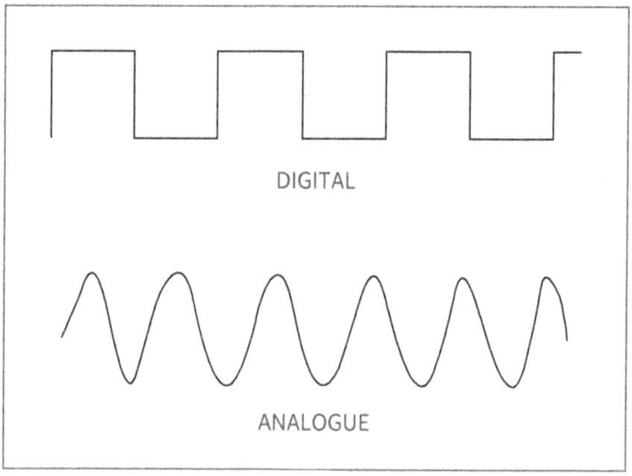

Figure 7.1 Digital data versus analogue waveforms.

There is a crucial reason for these distinctly different forms of information profiles between living systems and machines. All living organisms are sentient, cognitive organisms that must individually evaluate their information to determine both its quantitative and qualitative measuring value (Miller, 2018; Miller et al., 2020a). Indeed, it is this living action that specifically separates the animate from the inanimate. Living organisms have the capacity to discriminate information content so that physical, informational cues become meaningful biological information. Most importantly, living organisms are information dependent and actually participate in creating what that information means. Moreover, they communicate their qualitative appraisal of that information abundantly at all scales. For example, the cross communications among bacteria in biofilms to engage in quorum sensing is so active that it has been described as chatter (Visick & Fuqua, 2005). However, that communication represents an information exchange between 'knowing' cognitive agents and that 'knowing' reflects their sentient appraisal of both information quality and quantity. Obviously, this is different from computer data analysis since each of these cognitive players must assess information within their own context and limitations. Consequently, not only is the information that cells receive imprecise, but so is all the information they communicate to other living organisms.

There are several reasons that biological information is ambiguous by definition. For example, in living systems, neither the sender nor the receiver of information is necessarily known to one another. Substantial amounts of cell–cell communication represent a general broadcast, either as a profusion of bioactive molecules or bioelectrical signalling (Witzany, 2012). Further, any communication or environmental cues that a cell might receive must travel through an intervening medium, such as the cellular interstitial spaces, to reach the cell and deliver its information content.

That transit is a source of information noise because of inevitable degradations by the intervening media and time delays that are naturally imposed between the initiation of any information exchange and its reception (Miller, 2018). Moreover, all cells have an external membrane essential to their information system as a critical aspect of their senomic apparatus that connects them to the external environment (Baluška & Miller, 2018). Within this senomic chain, the cognitive cellular measuring assessment of information and its eventual contingent deployment directly relates to innumerable interrelationships within the complex, crowded environment of the cellular interior involving various organelles, each with their own membranes and the varied components of the cytoskeleton. All are vital links within the senomic apparatus and are essential to the process but inevitably contribute additional sources of noise.

Beyond these readily understandable physical sources of error in information evaluation among cells, another constitutive source directly relates to the fact that all living beings are self-referential. Self-referential status means that all organisms are both observers and participants in decision-making processes (Torday & Miller, 2020). Unlike computers, no two observers or participants are identical in their internal measuring assessment of information in living communications. Obviously, as two separate entities, they are not in precisely the same place at the same time. Consequently, they never receive identical information inputs. Nevertheless, they react similarly to these minor variances in information inputs since they are self-similar and are conditionally triggered to interact with one another. However, information quality takes on differing characteristics for each of them as separate observer/participants (Miller, 2018).

It is an important point that this observer/participant effect is not merely the result of their physical separation. Certainly, they both can not occupy the same space simultaneously. However, there is another pertinent factor at work in a self-referential framework. There are obliged differences in information perception and its consequent analysis that stem from variances in the assessment of thermodynamic variables and quantum phenomena which are all guided by inherent uncertainty relationships, codified as the *Heisenberg uncertainty principle*. In physics, it is not possible to know simultaneously both the position and speed of a particle, such as a photon or an electron, with perfect accuracy. The measurement of one variable changes the value of the other. Consequently, different observers will use their senses to evaluate environmental cues, some of which are quantum dependent. The result will be inter-observer variances in the assessment of speed, amplitude, or quantity of experienced phenomena. This principle applies to all aspects of the cell, including its genome (Strippoli et al., 2005). Even for our genes, considered by some biologists to rather closely reflect a digital system, there is never any possibility of knowing with absolute certainty what the exact composition of the genome is at any moment in time. As will be covered in Chapter 8, the genome is no longer regarded as a static repository as it once was. Instead, it is a dynamic reciprocating partner in cognitive cellular living processes with its own rate of inherent mutations, insertional mutations, self-generated editing, and epigenetic impacts. An evaluation at any moment is a snapshot of something in motion that does not represent an absolute.

Other important, embedded variances in the self-referential information assessment are unique to the observer–participant context. The most significant cause of this living effect is characterized as 'antipodal' information that offers insight into how dependent individual observer/participants are on their circumstances when evaluating environmental cues. Four examples of this type of antipodal information are worth examining.

The first type views information as a volume, like a sphere with dimensionality rather than a specific point that would convey higher measuring value. To conceive this, imagine two observers standing on either of a soccer ball tossed in the air. Both see a ball, but each sees a different side as it rises and falls in the air. For each, one side is visible, and the other is hidden. However, for each, what is 'en face' is 'opposite' or 'antipodal' compared to the other. Consequently, they might construe differences in meaning exactly what that soccer ball represents (Tozzi & Peters, 2017). For instance, there might be painted lettering on one side and not the other, and each would 'understand' the ball differently within their separate observer context. Same ball, different net self-referential appraisal.

The second type has been categorized as a 'distinction on the adjacents' (Marijuán et al., 2015). The term 'adjacents' references the obligatory gap between information reception, its intracellular measurement, and its deployment. Uncertainties in the measuring value of information can be narrowed by bringing 'adjacents' together. This scenario can be conceptualized by regarding information content as something like a dimple on a golf ball. Suppose the information content is imagined as the dimple. In that case, the uncertain spread of that information can be narrowed if the assessment proceeds across the diameter of the dimple rather than its circumference, or if the edges of the dimple can be brought closer together.

A third crucial factor in information assessment is easy to disregard but is still imperative. There is substantial measuring value to the gaps between 'bits' of information. Notably, these gaps as space between the formal information content are meaningful and have been dubbed a 'third state' of information (Forshaw, 2016). The third state contributes organizational structure to information just as pauses between words connote semantic information in speech or the spaces between the words in this sentence contribute to its ready understanding.

There is one further highly significant type of inter-observer informational variability that is often overlooked and specific to the self-referential frame. Cognitive organisms use information to predict and anticipate. Any absent but anticipated information is meaningful, as we all know in our lives (Miller et al., 2020a, 2020b). Significantly, this self-referential form of information is scale independent since it is experimentally confirmed that even single-celled microbes can anticipate and predict (Vallverdú et al., 2017).

From this background, we can now discuss two critical living realities driving biological and evolutionary development among conscious cells. The first is the presence of a definable information cycle that stimulates the cellular behaviours that permit multicellularity. The second is why the self-referential frame assures that any cell's information is exclusive to itself.

There Is a Crucial Cellular Information Cycle

From the preceding, where the quantity and quality of information constitute the working parameters of cellular consciousness, it is readily apparent that cellular life equates with information management (Miller, 2016a; Miller et al., 2020a, 2020b). When that is acknowledged, an understanding of the origin and persistence of multicellularity representing the dominant form of life on the planet emerges. That path lies within the mandated properties of the cellular information management system. Accordingly, given this overarching framework, it becomes apparent that information has crucial self-organizing properties in a self-referential framework (Miller, 2018).

As has been emphasized, all the information any cell has is conditioned on its imprecision. Consequently, all cells must measure any information they receive to assess its meaning. This prescribed process of the self-assessment of ambiguous information drives two key features of a biological information cycle. First, since all cellular cues are ambiguous for the reasons that have been enumerated, cells have learned that the measuring value of information can be improved by its collective measurement as the cellular expression of the 'wisdom of crowds'. Sentient organisms 'know' that their individual internal assessment of information is an imperfect model of their circumstances. How do sentient organisms cope? They do just as we do ourselves when feeling conflicted. We ask others for their opinions. This innate tendency is present at all living scales, including our own. This same obliged impulse forms the basis of many TV game shows. What do contestants do when asked to respond to a multiple choice question and have no clue about the answer? They appeal to the audience for their collective judgement. They even do the same for random selections. Do I choose door number 1, 2, or 3 to get the best prize? Cells do the same and exert the 'wisdom of cells' to improve their individual assessment of the validity of environmental cues. Humans and cells instinctively 'know' that they do better following collective judgements and obtain the 'wisdom of crowds' rather than relying solely on their own limited ability to assess information unless they're academics who delight in disagreeing with others, as we do.

Accordingly, the collective assessment of information improves its validity for cellular decision-making and enabling predictions as effective information (EI*) (Miller, 2018). Maximizing EI* directly links to improved energy efficiency since the decision tree towards cellular decision-making is strengthened and the deployment of its limited resources is improved. This aspect of information quality and how it relates to informational imprecision has been categorized as an 'uncertainty relation', which in data transmission refers to the number of independent data points available over time to make the analysis (Gabor, 1946). Simply put, more independent measurements by self-similar cells mean narrowing this uncertainty relationship and a higher measuring value for the information in question.

Importantly, the shared assessment of information is not merely a learned response by cells. The only requirement to set the cellular information cycle in motion is the reception of information by any self-referential organism as an observer/participant. Any information that is received must be measured, and that process of internal analysis is necessarily work. That work is an obliged consequence of the cellular internal measurement of information whose analysis necessarily requires an energy expenditure. That cellular work leaves an obligatory energetic signature relating to both the quality and quantity of the information being assessed. In turn, this activity becomes a communication to any other observer/participant within the same system as an energetic input. Moreover, that communication, itself, obliges that next observer/participant to engage in its own self-referential measuring assessment of that communications informational meaning. This also constitutes work, inevitably delivering a reiterating signature that becomes yet another communication. The result is an obligatory, reiterating work channel (Deacon, 2011). This specific obligatory feature of biological information analysed within a self-referential framework accounts for multicellularity (Figure 7.2). From the cycle, it becomes apparent why biological information is inherently self-organizing and all of biology is information management (Miller, 2018; Miller et al., 2020a).

All cells receive biological information from the environment characterized by its imprecisions, triggering an obligatory self-referential cellular information cycle. The reception of information requires its measured internal assessment to determine its value and meaning for potential communication and deployment. However, irrespective of that exact measuring value, any internal assessment involves an expenditure of energy which becomes an obligatory environmental cue to some other cell (observer/participant). Any cell receiving that cue is then obliged to assess it, stimulating a further work channel. Cells measure information together to improve its

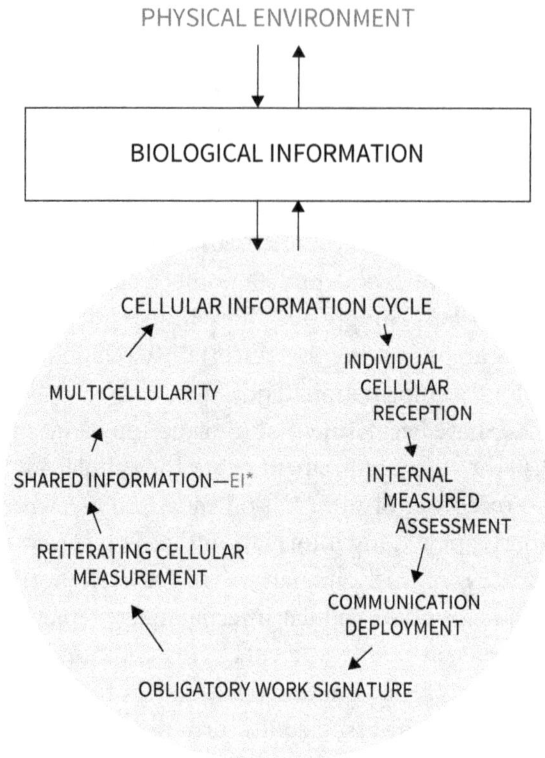

Figure 7.2 The cellular information cycle.

measuring value as EI*, which translates into energy efficiencies, the willingness to trade resources, improved cellular homeorhetic equipoise, and survival. The tightly woven, obligatory cellular information cycle explains why multicellularity is a dominating form of life on the planet, either as biofilms or holobionts.

This living imperative is no small matter. It explicitly accounts for all multicellularity. If cells did not have the 'knowing', sentient impulse to collaborate with other self-similar (recursive) organisms to strive towards better informational validity through collective measurement, you would not be here. How strong is this biological imperative? It accounts for the seamless integration of the tens of trillions of co-dependent microbes that partner with your personal eukaryotic differentiated cells, enabling you to be a successful reproductive holobiont. What does this mean in strictly biological terms? All cells, and derivatively all multicellular eukaryotes as holobionts, such as you, are an intricately integrated informational interactome.

Your Cells Create Their Own Reality—So Do You

The nature of cellular information assessment has a further crucial entailment critical to cellular life and yours. Both cells and we create our own realities. Although, at first consideration, this assertion might seem to lie within the purview of philosophy and metaphysics, or via a deep dive into consciousness studies, the correct analysis is directly rooted in our obliged biological context as completely cell-dependent self-referential organisms.

All individual self-referential, conscious agents are obliged to assess imprecise environmental cues through their individual tools for information assessment. Since the only information a cell can have about any external reality must travel across an outer membrane and be analysed internally, any information a cell has is internally generated. A cell receives an ambiguous environmental stimulus and measures it. That is its only real information, as the self-produced awareness of its measuring value based on its own self-determined assessment. That is the totality of the information that a cell has. A disturbing conclusion derives from this inescapable circumstance. There is no such thing as a completely objective measuring value of any physical information. All our information is our self-referential creation through self-production (Miller et al., 2020a). This living reality has been formalized in the concept of info-autopoiesis, which axiomatically shows that every bit of the information a cell possesses is the product of its internal self-creation (Cárdenas-García, 2020). Although cells of like type are self-similar, they are never identical. Consequently, each cell will have constructed its exclusive interpretation of its external milieu.

Since we are assemblages of trillions of cells, each producing its assessment of information content and quality, it is inescapable that we create our own derivative reality as a summation of our cellular information content. Naturally, this accounts for the wide range of variances that self-referential organisms, including ourselves, display towards environmental stresses. Of course, we intuit this principle since we all acknowledge that we do not feel each other's experiences. And further, who among us has not commented at least once about someone else as we have tried to fathom their contrary opinions or actions and said with a resigned shake of our heads, 'They just live in their own world'.

A significant feature of 'reality' to a cell is that its interpretation of the external environment must be conducted through the veil of its own internal noisy processes. One unquestionable aspect of our living circumstances is that we are surrounded by noise. That noise is received information that is instinctively regarded by us as part of a continuous background state, typically not rising to the level of requiring overt attention. Cells live in a crowded, active milieu with a flood of molecular and energetic flows. It is mistaken to assume that a cell only depends on high-amplitude information inputs. Indeed, it is quite the opposite. Cells rely on all types of inputs, including the continuous monitoring of their background noise as a further cue to its equipoise. For example, consider the background, noisy chatter in the jungle. If you were travelling through it and it was suddenly interrupted, you might interpret that as consequential information. Perhaps a predator lurks. True, the general run of these noises seems random, however, there are characteristics of that type of information that can be put to constructive use. Similarly, when we engineer, some chance events are seen as a serendipitous boon, helping to solve a problem. This concept of noise as a useful contributor to biological information assessment, intermittently representing useful information has been formalized as the concept of the 'harnessing of stochasticity' (Noble & Noble, 2018). Notably, this is not merely a theoretical concept. Research experiments in habituation in plants, animals, and even single-celled organisms confirm that the concept of the 'stochasticity of information' is real and context dependent (Eisenstein & Eisenstein, 2006; Gagliano et al., 2014).

Since it is insisted that life is an informational interactome, articulating a unified theory of integrated information that can be applied to biology is desirable, beyond the categorical acknowledgement of its ambiguity. The relatively recently introduced concept of integrated information theory (ITT) can be productively applied to the patterns of biological information and consciousness studies since it focuses on 'nested discriminations' (choices) (Tononi, 2008). In particular, ITT is an attempt to better explain the contentious issue of qualia (individual and specific subjective instances of conscious experience, e.g. the taste of wine). In ITT, qualia as subjective experiences represents integrated information that is internally generated (Tononi, 2004).

Fortunately, the major tenets of ITT apply directly to a cellular approach to consciousness, placing it on a solid footing of information reception and analysis (Oizumi et al., 2014). In ITT, conscious is structured. Further, every instance of an experience is a combination of inherently integrated aspects such that it cannot be reduced to its individual components. Naturally, consciousness is also informative. And each instance of an informative experience is exclusive and has separate particulars compared to all others. All of these particular facets of ITT relate to cells and their obligatory framework of self-produced information.

What can we glean from this approach to cells and their information-dependent circumstances? The reason our experiences are individual and idiosyncratically our own is because it is the same for cells, and we are cellular beings. Consequently, a theory of mind or consciousness is a cell theory. And further, every stimulus that impacts the cellular senome constitutes a cellular experience since all such impacts are highly integrated across the cell. Consequently, any attempt at undue reductionism, taken outside of the cell as an 'organized whole', will lead to a fundamental misunderstanding about the enigma of consciousness. Albert Szent-György, a Nobel Prize-winner in Physiology or Medicine, vividly grasped this principle, stating, 'My own scientific career was a descent from higher to lower dimension, led by a desire to understand life. I went from animals to cells, from cells to bacteria, from bacteria to molecules.... On my way, life ran out between my fingers' (as quoted in Gómez-Márquez, 2020).

Cells as Biological Expressions of an Information Architecture

When we regard our consciousness, it is apparent that we somehow integrate the myriad environmental cues that our senses provide us to form our experiences and enable our decisions. Accordingly, the concept that our consciousness is some form of information management seems intuitively obvious. It is the same for our cells. Their information management system is based on the exquisite links across their cellular senome, which enables the cellular measurement of information (Baluška & Miller, 2018). There is a direct reason for this complex set of connections. Cells have a senomic information architecture since it governs their self-production of information, enabling cellular prediction as inference and anticipation. That architecture seamlessly connects cellular problem-solving and triggers the cellular communication of information appraisals to improve the critical deployment of cellular limited resources as bioactive molecules or energy. Consequently, self-referential cells have a cellular information architecture as a crucial component of its self-referential information management system (Miller, 2016a). Our human capacity for abstract thought and greater engineering prowess must therefore relate to our emergent ability to juggle more informational uncertainties prior to our contingent decisions to deploy our own limited energetic resources.

Our cell-based self-referential information management system is sustained by a fundamental fulcrum. In a self-referential framework, energy as environmental cues represents self-referential information which obliges communication. Thus, our world of biological expression, which is exclusively cell-based, relies on triadic energy–information–communication to propel biological expression (Miller, 2018). Crucially, that triad links to an impelled work channel. Multicellularity is the biological expression of that work channel, whose characteristics will be further described in Chapter 8 that explains the critical patterns of biological and evolutionary development that have contributed to all planetary life.

When information assessment is appraised as the epicentre of cellular consciousness, the intricately coordinated cellular confrontations with environmental stresses can be appreciated as indicative of a highly attuned information management system, permitting the orderly assessment, measurement, communication, and deployment of information by self-referential cognitive cells. What do cells know? They know that their information is imperfect and they know to seek others to share information to improve its quality. Intimately linked to that is the principle of recursion that grants the knowing of self-similarity, as the sense of 'as here so elsewhere' (Kafatos, 2014). By this means, self-similar, self-referential organisms react to stimuli in generally concordant patterns, that is, they know together.

Consequently, cells 'know that they know', and 'know that others know' (Miller et al., 2020a). This is the informational bioactive substrate of multicellularity that depends on shared information. In this way, self-similar, self-referential organisms constitutively self-produce their own information, but can collaboratively cooperate since they share a binding informational motif. They may not be identical, but they are of like kind. This informational nexus is the glue that binds multicellular organisms.

How does ambiguous information centre in this narrative? It exists at its epicentre. The realization that biomolecular processes represent informational processes is growing (Miller, 2016a; Baluška & Miller, 2018; Miller et al., 2020a, 2020b; Fields & Levin, 2021; Marijuán & Navarro, 2022). Consequently, the flow of information is now regarded as central to all biological processes. Pertinently though, information flow in biological contexts mirrors energy flow (Miller, 2018; Marijuán & Navarro, 2022). Just as energy must be channelled to do productive work, the trick in living systems is to control the appropriate flow of information to permit contingent cellular problem-solving. Accordingly, the characteristics of information in biological processes are paramount to our understanding, which can be ascribed to two salient attributes. First, as stressed in this chapter, all the information that any organism has about itself is characterized by imprecisions. And second, the heart of living information management governing its productive flow is cellular sentience.

Within the foregoing, two further conclusions directly ensue. The objective of cellular measurement is attaining and sustaining a preferential state of homeorhetic equipoise in continuous confrontation with environmental variables. However, and unequivocally, the cellular sentient appraisal of its equipoise is representative of its consciousness. Hence, the primary objective of cellular measurement clarifies. The cellular information management system is always directed towards preserving its instantiated consciousness within its embodied cellular form (Miller et al., 2019). The specific objective of the cellular measurement of information is to protect its self-referential integrity.

From an information perspective, uncertainty equates with incomplete information about any relevant information-dependent system (Khrennikov, 2007). That uncertainty is diminished through collective cellular judgement. Necessarily, cognitive faculties lie at the centre of all discriminating actions. How might that cognitive capacity of cells be characterized? Self-referential cells understand that the quantitative and qualitative aspects of their informational palette have constraints. Necessarily then, cells are measuring instruments (Miller et al., 2020a). Further, as environmental flux is life's condition, all cellular deployments of assets entail prediction and anticipation. Recent research confirms that single cells can do internal calculations and predict (Luczak & Kubo, 2021). A direct correlation has been found between the activity of individual neurons and their ability to accurately predict the validity of informational inputs, thereby minimizing 'surprise' which directly relates to energy expenditure. Modelling suggests that the exchange of predictions among individual neurons to aggregated cellular networks that realize more complex learned predictive patterns, ultimately representing human consciousness (Luczak et al., 2022). Moreover, pre-existing cellular states and multimodal perceptions shape cellular decision-making according to their internal states and external conditions (Kramer et al., 2022).

Less obviously, these same principles hold true for cellular communications (Miller et al., 2020a). It has been previously noted that any energy expenditure by a cell becomes a form of communication to others surrounding it. However, beyond that obligatory type of communication, cells communicate abundantly on a volitional basis so that they can collaborate. However, each such communication requires an energy expenditure. Accordingly, the cellular decision to expend that energy is a prediction. Is it worth the energy expenditure? All biological and evolutionary development must thereby represent the entanglement between the real-time cellular ability to assess conflicting environmental cues to resolve informational ambiguities and productively deploy cellular resources, either on its own or through collective action. That explicit gap between cellular environmental assessment of available information and its deployment to constructively meet environmental imperatives represents cellular prediction and explains the compulsory role of an overarching cellular information management system to efficiently manage its scant resources.

Frankly, none of this is conjectural. If cellular information were perfect, it would not require measurement, and consequently, a cellular informational architecture or a cell-centred information management system would be an unnecessary encumbrance. Perfect information requires no measuring assessment or any information management apparatus. Instead, cells leverage imperfect knowledge to sustain homeorhetic equipoise. Significantly, then, the collaboration, cooperation, co-dependence, and competition that characterize all multicellularity are the biological expressions of information flow. The driver of that flow of information is cellular consciousness that enables the productive predictions which sustain life (Miller et al., 2020a). Saliently, this accurate perception of cellular dynamics effectively reduces to a single encompassing biological and evolutionary narrative. Information-dependent cells dwell in self-referential doubt, and that incertitude drives our living world.

8

Genes Are Tools of Intelligent Cells

Biological and Evolutionary Development in the 21st Century

Introduction

This book is premised on the copious research demonstrating that all cells are intelligent, problem-solving agents (Ford, 2004, 2009, 2017; Shapiro, 2007, 2021; Lyon, 2015; Miller, 2016a; Reber, 2019; Baluška et al., 2021b, 2022b). As discussed in Chapter 3, the living context of cells is their active management of information. Cells assess information internally and, dependent on that self-generated appraisal, communicate to other cells, deploying their limited assets to sustain cellular balance and protect crucial self-identity. Cellular information assessment is dependent on the complex linkages that comprise the cellular senomic apparatus, beginning with the cell's plasma membrane in immediate contact with the external environment and extending across its entire internal milieu (Baluška & Miller, 2018; Miller et al., 2020a; Baluška et al., 2021b). Crucially, that internal assessment of information constitutes a measurement. Its *meaning* is an at-the-moment evaluation of the impact of that external environmental stimulus vis-à-vis the cell's self-referential state of preference that references to its robust retrievable and deployable memory. In simplest terms, any informational inputs (temperature, light, bioactive molecules) can be considered a deflection from the mean, which the cell measures to determine its contingent deployment of resources to meet this variable. As has been emphasized, cells measure information since all cellular information is imprecise. Perfect information does not require the expenditure of scant energy for its assessment.

Evolutionary biologists have traditionally overlooked that cells measure information; indeed, the study of evolution has not traditionally centred on cells. Instead, it began as a discipline devoted to rigorous observation and, in the latter half of the 20th century, devolved into a focus on population genetics. Consequently, the ramifications of the critical issue of the ability of intelligent cells to measure information was unappreciated until the last few

The Sentient Cell. Arthur S. Reber, František Baluška and William B. Miller Jr., Oxford University Press.
 DOI: 10.1093/oso/9780198873211.003.0008

years (Miller, 2016a, 2018; Miller et al., 2019, 2020a, 2020b). However, the scope of cellular measurement of information is vital to all biological and evolutionary development.

In this chapter, the direct connections between measurement, communication, and natural cellular engineering will be investigated. Through that exploration, it will be explained why the prior concepts of genes as the centre of biological activity must be thoroughly reconsidered in a modern context. In 20th-century neo-Darwinism, genes were presumed to control cells and determine cell fates. In the 21st century, genes are acknowledged as tools of intelligent, measuring cells. Instead of the prior conception of our central genome as a nearly inviolable genetic repository, it has been conclusively demonstrated that genes and the entire genetic panoply of cells are flexible and malleable participants in cellular problem-solving.

Our Contemporary View of Genes

Any re-examination of the concept of the role of the gene in biological and evolutionary development has a corresponding derivative. The acclaimed concept of Crick's central dogma must be freely reappraised. Crick's central dogma of molecular biology states that the information stream in cells follows a strict directional path from DNA to RNA to proteins. That interpretation of biological molecular flow strongly encouraged a belief system asserting the primacy of genes. Recently, major deficiencies in that dictum have become apparent. For example, the reverse transcription of DNA from RNA was once believed to be rare. Recent research now suggests otherwise. A theta polymerase repairs DNA converting RNA code into DNA in cells as efficiently as HIV reverse transcriptase (Chandramouly et al., 2021). Moreover, the unidirectionality of the flow of cellular information from DNA to RNA to proteins is also contravened by two specific factors whose impact is now being more fully explored. First, the extent and heritability of epigenetic changes had been vastly underestimated in the past. Epigenetic changes are now known to significantly affect the course of evolution as a source of genomic plasticity and modulation (Torday & Miller, 2020; Ashe et al., 2021). Second, organisms continuously adjust their genomes through read–write genomic editing (Shapiro, 1992, 2019; Witzany, 2011).

The originating conceptual frame underlying modern epigenetics is generally accredited to Jean Baptiste Lamarck in his 1809 *Philosophie Zoologique*. Lamarck conceived of an evolutionary system where individuals developed new traits by reacting to their environment, thereby acquiring heritable characteristics. A frequent example of a Lamarckian type of inheritance would be the theory that giraffes acquired their long necks by constantly reaching for leaves in trees. Although initially rather popular, it became identified with vitalistic thinking and was ultimately discredited since no specific biological mechanism could be identified in earlier decades.

However, in a series of inventive experiments with *Drosophila melanogaster*, Waddington (1961) seemed to confirm the possibility of external forces altering gene expression through a process of 'genetic assimilation'. Environmental stresses could trigger biomolecular processes that regulate genetic expression independently of the specific DNA sequences. Waddington termed this process *epigenetics*. Such epigenetic changes have been confirmed as a source of heritable variations through processes such as methylation that modify DNA structure while leaving the underlying code unchanged. Other epigenetic changes can result from horizontal gene transfers, which include transposable elements (TEs), interspecific hybridization, viral incursions, or shifting relationships with a contributory microbiome from symbiosis or parasitism (Jablonka & Lamb, 2008; Torday & Miller, 2020).

A variety of epigenetic influences affect the nuclear genome. Accordingly, heredity is now understood to be a much more pluralistic process than was originally thought. In the 21st century, the genome is now considered a reciprocating organ within an organism as an organized whole. Since phenotype is expressed through the proteome, even though the proteome may not directly alter DNA, its information content distinctly reverberates back to the genome to affect its regulation and expression (Burley & Kamada, 2002; Stadhouders et al., 2019).

Contrary to long-held presumptions, the current view of the genome regards it as a highly flexible, adaptive, read–write informational system that directly and responsively contributes to evolutionary innovation (Shapiro, 2013). The genomic read–write system permits cells to actively write their information content into their genomes through natural genetic engineering, epigenetic formatting, structural adjustments, symbiogenetic cell mergers, horizontal transfers, and interspecific hybridizations (Shapiro, 2013, 2017). The exact details of these complex processes need not be our specific concern in this chapter. However, it is important to mention these as they indicate the wide range of processes that can flexibly alter genomic content. Further, another type of genetic input, termed mobile genetic elements, disperse across genomes, representing pertinent information that functions akin to 'plug-in' cassettes. These small snippets of genetic code can modify cellular networks by impacting nucleoprotein structures and, through domain shuffling, ultimately affecting phenotype (Shapiro, 2016). Reciprocally, phenotype affects the genome, forming flexible, adaptive pathways to meet contemporary environmental stresses. The extent of these interchanges is profound. It is currently estimated that mobile elements comprise the majority of eukaryotic genomes (Arkhipova & Yushenova, 2019), and further, these insertions are not random (Miller et al., 2021). Indeed, there is growing evidence that a substantial proportion of gene editing is non-random (Zamai, 2020).

Accordingly, TEs, additional types of genetic particles called integrative and conjugative elements, and the great varieties of small RNAs all constitute an environmentally responsive organism-wide, multi-domain genetic mobilome. These genetic players represent an 'on-call' information repository for the flexible, adaptive responses of cells to environmental challenges (Miller, 2016a; Miller et al., 2021). Necessarily then, evolution must extend beyond its prior genomic focus to accommodate the entire panoply of sources of information that can affect cells, their genomes, and other genetic constituents. All of this active genetic mobilome participates in finely tuned coordinated processes, incurring viruses, and competent RNAs (Witzany, 2009, 2014). And further, all of these genetic constituents are elements of the read–write informational system of cells that interact with its essential internal organelles and plasma membrane.

The proteome, too, is undergoing its own reconceptualization. Indeed, the proteome is now understood to be more complex than either the genome or transcriptome (Carbonara et al., 2021). Rather than representing canonical proteins, the proteome is being re-imagined as a myriad of 'proteoforms' emerging as isoforms, splice variants, and separate modifications before, during, or after transcription and translation (Carbonara et al., 2021). There is a growing appreciation that proteomics reflects the actual status of the cell compared to genomics or transcriptomics. This accumulating knowledge directly influences our understanding of cell dynamics, especially in immunology and cancer research (Urbiola-Salvador et al., 2022).

Over the last several decades, research has revealed the substantial epigenetic evolutionary impact of TEs (Warren et al., 2015). TEs were originally considered little more than genetic parasites capable of transmitting some genetic information (Hua-Van et al., 2011). However, their contribution to biological and evolutionary development and an extensive range of activities has been clarified (Federoff, 2012; Hedges & Belancio, 2011). TEs as 'jumping genes' are small segments of 'restless' DNA that can shift position in genomes, affecting its size or range of expression. Some of these remain within an ancestral genome, while others can be transmitted across species boundaries where they can be either active or dormant (Hedges & Belancio, 2011). That interspecies transmission relates to the fact that several TEs have been found in viruses that specifically target eukaryotes. These can act as vectors of horizontal transmission to eukaryotes as a form of infectious spread. Since TEs can affect gene regulatory networks and either enhance or silence adjacent genes, it is not surprising that scrupulous intracellular mechanisms exist to regulate TE expression as part of the information management system of cells.

At one time, it was assumed that gene duplications were responsible for the size of genomes. Consequently, it was surprising to discover that up to two-thirds of the human genome consists of TEs. These TEs not only expand genomic size but stimulate further genetic rearrangements, ultimately affecting genetic expression in both animals and plants (Hirsch & Springer, 2017; Drongitis et al., 2019). Some TEs, such as LINE1, repeat 500,000 times in the human genome. For years, this component of our genome was thought to be 'junk' DNA. However, research has uncovered that LINE1 is an important co-factor in embryonic development and affects RNA networks (Fadloun et al., 2013). Further, TEs convey many transcription factors that influence molecular linkages in the genome but also extend beyond it to modulate immunity and cellular responses to environmental stresses (Wagner & Lynch, 2010). Accordingly, the modern picture of TEs emphasizes their substantial role in gene regulation and expression. Altogether, these various genetic factors contribute to our modern understanding of the central genome as a flexible and even malleable genetic repository that contributes to biological variation and evolutionary diversification.

Among the most evolutionarily significant TEs are infectious retroviruses, another form of retroelement (Boeke & Stoye, 1997). Retroviruses have helped shape evolutionary outcomes, contributing adaptive value (Shapiro, 2016). For example, retroviral insertions account for the origin of syncytiotrophoblastic tissues of the mammalian placenta (Black et al., 2010). Indeed, the substantial array of retroelements in species' genomes can be seen as residues of previous infection interchanges across evolutionary development, with many of them contributing phenotypic and metabolic variation and serving as a source of biological novelty (Witzany, 2011; Bennetzen & Wang, 2014; Soucy et al., 2015).

The discovery of all the preceding means of adding or adjusting the genetic complement of cells and organisms commissions an enlarged framework of the true nature of our genetic complement. Genes are tools of flexible cells functioning as retrievable and deployable memory. Genes serve sentient, competent cells, enabling accurate self-similar reproduction and permitting biological variations as flexible problem-solving to meet environmental stresses.

An Engineering Cycle Derives From the Information Cycle

Because cells measure information, they communicate. If information were perfect and measurement was not required, communication would not be needed. The point of communication is to convey the meaning of the self-produced internal assessment of information to other cells. Communication is the sharing of information, and as previously indicated, this translates into improved validity in the internally measured assessment of ambiguous environmental variables through collective appraisal. This aggregate information assessment is termed effective information (EI*) since it represents the summation of multiple separate cellular information measurements, which propels multicellularity (Miller, 2018). Further, as introduced in Chapter 7, any internal assessment information is an expenditure of work. That work issues an obligatory work signature, leading to a reiterating information cycle (see Figure 7.2 in Chapter 7).

Any reiterating cycle in biology that is energy expensive has a purpose. Undoubtedly, any sharing of information is an expression of cellular cooperation and an evident distribution of resources. Necessarily, this collaborative impulse is an active expression of cellular consciousness. Although the cellular information cycle involves some obligatory cell–cell communication, the sharing of the internal appraisal of information as measurement is predominately contingent on cognitive assessment. Pertinently, however, it is not just that cellular consciousness permits the cooperative sharing of information. All life depends on this living feature. Even unicellular organisms such as bacteria freely and abundantly associate via highly integrated and complex, collaborative biofilms as a vast consortium of different unicellular genotypes and phenotypes (Wojciech et al., 2018). Consequently, it is not just that cooperation can be attributed to cellular consciousness. Cellular consciousness is explicitly dependent on cooperation to continue to exist. The perpetuation of the cellular form, which equates with the perpetuation of self-identity, is dependent on cooperation, which explains the dominance of multicellularity on the planet. Correspondingly, all cellular processes are purposed for the continuous protection of cellular self-identity, which is its state of consciousness self-awareness (Miller et al., 2019). Consequently, cellular collaboration is not an epiphenomenon of consciousness but is an embedded living attribute, inseparable from its entirety.

How does this obligatory interrelationship express further in biological terms? Because cells can measure and communicate, they can engineer. Cells cooperate to engineer collective solutions to cellular stresses, a process that has been described as natural cellular engineering (Miller et al., 2020a, 2020b; Torday & Miller, 2020). What are cells engineering? They are engineering conducive habitats as niche constructions. For example, the localized tissue ecologies that are part of your body, like your intestines and gut microbiome, are examples of natural cellular engineering as niche constructions. You are their product, and so is every other visible creature on the planet (Miller et al., 2019, 2020a). All organisms are part of a reiterative cellular information cycle (Figure 8.1).

The external cellular environment offers biological information as cues, triggering the cellular information cycle. The dynamics of the cellular information cycle stimulate a reiterating work channel based on the collective appraisal of information (EI*), deployed among cells as collaborative natural

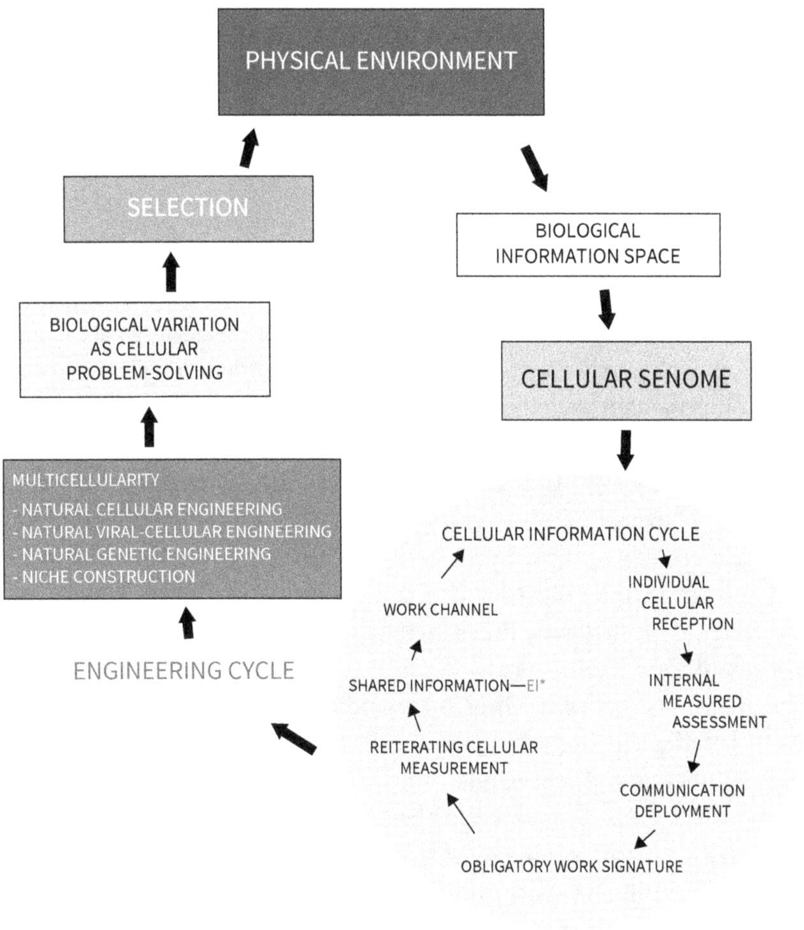

Figure 8.1 The cellular engineering cycle.

cellular engineering, which can include a co-partnering viral contribution. This coordinated cellular process produces unicellular collective biofilms and the united tissue ecologies that constitute all multicellular eukaryotes. Biological variations result from differing natural cellular engineering outputs as differential cellular expressions of cell-based problem-solving. Those cellular solutions that are 'fit enough to survive' continue the reproductive cycle.

How do we humans engineer? We appraise information, communicate, cooperate, serve co-dependent functions, and compete. What are we engineering? Solutions to human problems and concerns, such as houses, cars, jet engines, books, scotch tape, or air conditioning. All are solutions to problems or, alternatively, attempts to achieve a state of preference, which is just another form of problem-solving. Quite directly, the reason that humans can engineer is that their cells do. We are their engineering product. If they couldn't engineer, we would not be able to do so, because we would not be here. How do we engineer? We use tools, like hammers or saws. Cells use their tools. All aspects of the cell are its tools. Indeed, it has been pointed out that the cell itself should be regarded as the first example of niche construction (Torday, 2016). This niche construction activity that created the crowded, active environment of the cell is a masterpiece of internal engineering.

Humans use tools of all kinds and depend on retrievable and deployable memory to conduct successful engineering operations. All the vital components of the cell are its tools, including its genes. Accordingly, the senome, genome, and epigenome are all tools of the intelligent, measuring cell. Even the genes in our obligatory constituent microbiomes serve our cells. Genes are a crucial component of the retrievable and deployable memory system of cells. This perspective is quite the opposite of what had been previously believed. In 20th century biology, genes were the masters and controllers of cells. Cells exist to serve genes and their perpetuation. Despite this confusion, which was most effectively fostered by Richard Dawkin's celebrated *The Selfish Gene* (1976) and came to be regarded as the most influential science book of the 20th century, the conclusion that genes are tools of cells is intuitively obvious. Genes outside of cells are inert.

Exactly how are genes serving? Genes aid consciousness as part of cellular problem-solving. In particular, genes serve by being critical tools in natural cellular engineering. There is one other obvious requirement for any engineering project. Cells, as engineering participants, require a plan or some type of template. Although it is in vogue to insist that this plan simply emerges *de novo* from the conjoining action of cells, self-produced electrical gradients, and bioactive communications (Hao et al., 2021), an actual informational template is required (Miller et al., 2020b). Although genes are tools of that templating process, the actual plan lies in an overlying information matrix, which will be discussed in the next chapter. When genes are properly placed into their context as flexible tools of cells that use them to solve problems, five further implications derive, some of which are not obvious.

1. First, cells are engaged in collaborative engineering. Consequently, this collaborative engineering produces the complex cellular ecologies that constitute the bodies of all multicellular eukaryotes (including you). All phenotypes are the result of concordant natural cellular engineering (Miller et al., 2020a). Notably, phenotype not only includes observable features like height, weight, hair colour or texture, or limbs but all metabolic processes. If you look up the definition of phenotype, the internet will commonly define phenotype as genetic expression. However, there is a vital difference that should be fully accounted for in modern biology. That genetic expression is a reciprocating effect of the cellular assessment of environmental cues and collaborative cellular decisions about how to productively adjust to new stresses. Consequently, phenotypic variations are the result of concordant differential natural cellular engineering and niche constructions in reciprocation with a flexible read–write cellular genetic complement.
2. When coordinated biological expression as phenotype is accredited as being due to natural cellular engineering, it is unquestionably driven by largely non-random processes. That coordinated engineering is a direct manifestation of cellular problem-solving in response to environmental and epigenetic impacts (Miller et al., 2021).
3. Since all multicellular eukaryotes are holobionts with active participants from each of the four domains (Prokaryota, Archaea, Eukaryota, Virome), any holobionic multicellular variation as co-engineering must be a product of co-engineering among representatives of all four domains (Miller et al., 2020a). Consequently, the nature of the relationship of the virome to cells requires re-evaluation. Although viruses are typically described in terms of host–pathogen interactions, that represents a small minority of the complex interplay between viruses and cells. In the main, viruses are co-partners with cells in natural viral–cellular engineering, working along with cells in maintaining suitable niche constructions.
4. Natural cellular engineering and its coupled process of natural genetic engineering are expressions of cellular problem-solving. However, the explicit purpose of cellular problem-solving is precisely directed to the protection of the self-integrity of all cellular constituents (Miller et al.,

2019). Cells collaborate volitionally, and cells have learned that their individual interests are best served within a collaborative form. Indeed, this individual cellular impulse to protect both individual self-identity and the self-integrity of others is so ingrained that it has been characterized as form of altruism. As noted in Chapter 2, Gürol Süel and colleagues at the University of California, San Diego found that distinctly separate colonies of *Bacillus subtilis* will communicate distress to one another, stimulating volitional reciprocating adjustments in nutrient uptake of the non-stressed colony in response to nutrient depletion in a common environment (Prindle et al., 2015). Cells in one colony will sacrifice to aid the other.

5. When genes are acknowledged as tools of cells subordinate to and reciprocating within a replete cognitive cell-based informational interactome, the general narrative of evolutionary development requires a complete reappraisal. In sum, and as will be discussed next, cellular consciousness changes all previous theories of biological and evolutionary development. Almost everything that comprises the core of the modern synthesis requires revision since the narrative changes from genetic mutations and gene frequencies to our contemporary understanding of intelligent, measuring cells as organized wholes.

Biological and Evolutionary Consequences of Genes as Tools

Darwin's 1859 *On the Origin of Species* continues to be acknowledged as the seminal text in evolutionary biology. However, it is often overlooked that his inspiration was a theory of forms (Bowler, 2003). With the advent of genetics in the mid-20th century, evolutionary biology was transformed into a theory of genes, becoming instilled within the prevailing, canonical neo-Darwinian modern synthesis. In Darwin's time, two central questions dominated the debate. How can life's diversity and history be explained, and how do form and function match (Pigliucci, 2007). Lamarck offered an answer through continuous environmentally induced adaptation, and his conjectures were well regarded in the 19th century (Gould, 2002). Indeed, it is often overlooked that Darwin was partial to Lamarckian precepts, offering his own version, termed 'pangenesis' in his 1868 *Variation in Plants and Animals under Domestication*, attempting to justify the transfer of traits of somatic cells to offspring under environmental pressure. Despite summary rejection during the 20th century,

Lamarck's perceptions are now reinvigorated within the burgeoning science of epigenetics.

Similarly, other previously sacrosanct areas of evolutionary biology are being substantially re-evaluated. Noble (2021) insists that the modern synthesis, first formulated in 1942 with the birth of genetic studies, remains steeped within four governing illusions: the absolute primacy of natural selection, the inviolable separation between somatic and germ cells, Crick's central dogma of a unidirectional flow of genetic information, and a rejection of Darwin's gemmules which presumed active cross-connections between somatic cells and the germ line. All have been contradicted by modern studies, but stubborn beliefs have impeded the development of a fully coherent alternative to the modern synthesis.

Of these outdated concepts, perhaps the most nettlesome has been the commanding influence of Crick's central dogma asserting primacy of genes. Nonetheless, the accuracy of that dictum has been increasingly questioned (Shapiro, 2009, 2021; Camacho, 2019). An embedded belief still exists that the central dogma remains true according to Crick's original pronouncement that 'once sequential information has passed into protein it cannot get out again', known informally as the 'sequence hypothesis'. Recent CRISPR analysis has begun to call even that into question (Ille et al., 2022). Nonetheless, any narrow vision of genetic adjustments misses the essential dynamic of a whole-cell flow of information that is certainly not unidirectional. That entire flow of information enables a cell to uphold itself, expressed as its self-referential self-integrity, sustaining its crucial homeorhetic balance and granting it adaptive flexibility (Miller et al., 2020a). Consequently, the genome and the rest of the cell-based genetic complement are considered facets of a whole-cell read–write informational architecture as a crucial component of the cell-wide critical feature of its information management system.

An alternative conceptual model is now offered that better explains the elusive nature of living organisms and the complexities of cellular dynamics. Specifically, the central dogma is replaced by a cell-wide read–write information system embodied within the cellular form that is compatible with framing biological and evolutionary development through the flow of information. Consequently, the cell must be regarded as an organized whole in which the proteome has a vital reciprocating impact on the genetic resources of the cell, contravening Crick's central dogma. Consequently, genes and genetic material are tools within the information management system of competent, sentient cells, enabling the requisite harnessing of stochastic and non-stochastic informational inputs to yield productive outputs. Through this differing frame, evolutionary biology is returned to its Darwinian roots as a theory of forms and not genes. Crucially, those forms are shaped by the congruent action of the entire suite of capacities of intelligent cells as organized wholes rather than genes as 'selfish' quasi-autonomous agents, as previously presumed.

Fortunately, contemporary research has spurred a substantial reappraisal of genomes and how the entire suite of genetic components in cells operate. That research has displaced the previous conception of genomes as a blueprint (Ball, 2016; Miller, 2018; Miller et al., 2020b; Torday & Miller, 2020). Previously, a common metaphor for the genome was that it represented 'the book of life'. However, it is now understood that the genome is not a book in any traditional sense. Instead, genes are a type of 'translational dictionary between two different worlds (languages), i.e. the world of nucleic acids and the world of proteins' (Cartwright et al., 2016). The now-refuted metaphor failed to account for the dynamic, read–write capabilities of the genome by which it flexibly serves continuous cellular adaptation. Further, the vital role of the large fraction of the central genome that is non-coding can now be honoured. Research has demonstrated that this extensive peripheral non-coding DNA protects the eukaryotic genome through a complex set of interactions. Hence, rather than a book, a genome is better analogized to an insect colony, which functions as an extended social habitat exhibiting common defence mechanisms (Qiu et al., 2017).

Consequently, 21st-century biology views the genome as speaking a common language deployed across the living planet (Torday & Miller, 2020). The genome and the entire array of genetic participants in cells play a crucial role in evolutionary development as one of the primary mechanisms of information transfer as community-wide cell–cell communication. Far from the common conception of viruses painted in neo-Darwinism within a traditional host–pathogen model, the virome is a critical co-partner in information transfer as cell–cell communication and as a co-engineering participant with cells in cellular problem-solving.

Therefore, viruses, retroviruses, subviral particles, and retroelements assist cells in their continuous adaptation to a shifting environment. Goldenfeld and Woese (2007) insisted that the neo-Darwinian conception of the virome as mere pathogens was a fundamental misapprehension. Instead, the virome should be viewed as a collective memory repository of community-wide genetic information, thereby contributing to evolutionary dynamics. Notably, viral incursions do not necessarily trigger immediate results as potential cellular co-partners. Both non-pathogenic and pathogenic viruses can display latency and sit quietly within the cellular milieu. In this sense, the virome can be considered 'on call' for future deployment in times of environmental stress, with resulting bioactive expression. Similarly, TEs, alongside retroviral inclusions, are now known to be appreciable participants in evolutionary development, partly accounting for continuous cellular adaptation but also intermittently triggering bursts of rapid speciation and evolutionary transitions that account for brisk adaptive variation (Oliver & Greene, 2011).

Cells can produce new phenotypes through collaborative natural cellular engineering. This process is the biological expression of multicellular assessment and measurement of informational cues of imposed environmental stresses. Through their collective action, congruent phenotypic variations as cellular niche constructions emerge. Necessarily, both random inputs and non-random cell-generated responses impact that coordinated cellular engineering (Miller et al., 2020a, 2020b, 2021). As part of this interleaved process, a logical processor decides if any generated new state is a preferable environmental response or should be discarded. That logic processor is cellular cognition/sentience.

At one time, it was presumed that the central genome subsumed this role as the controlling agency of cell fates. Modern studies indicate that the genome is crucial, but as a participant in a cell-wide read–write memory system rather than a developmental agency. Instead, the information management system of the cell permits the productive separation of conflicting environmental stimuli, distinguishing between random noise and significant cellular information, and melding them into integrated, purposeful outputs. Consequently, random inputs are channelled into practical biological expression as the 'harnessing of stochasticity' (Noble & Noble, 2018). In this manner, random events can be put to productive purposes. However, for that to be the case, there must be a biological logic processor coordinating this cell-wide information management system (Miller, 2018). Unquestionably, genes do not suffice. Instead, they serve the logic processor, which is, the cellular consciousness assumed in the CBC theory.

The old-fashioned model of evolution ensconced in the neo-Darwinian modern synthesis regarded random replication errors as the actual drivers of evolution (Bowler, 2003). Nonetheless, replication errors are subject to highly controlled error correction mechanisms in which the ability to correct errors via DNA polymerases can vary substantially depending on genome location and type of error (Kunkel, 2009). Epigenetic marks are also subject to revision, suppression, or deletion during two life-cycle stages. The cellular judgement of their biological suitability is largely determined during the obligatory recapitulation through the unicellular zygotic state and then further during early embryogenesis (Torday & Miller, 2016a). During this critical transition period, cells make decisions for their immediate and long-term benefits. Many of those epigenetic imprints are random occurrences. Nonetheless, some are differentially preserved and heritable. That differential has a significant impact on evolutionary development as a harnessing of stochasticity to sustain cellular balance or offer novel, beneficial results.

This same narrative determines any viral contribution to adaptive cellular responses. Viral genetic memory, either DNA or RNA, can be modified by several variational mechanisms, such as recombination, random point mutations, post-replicative repair, regulation of lysis, or relaxing polymerase proofreading (Sanjuan & Domingo-Calap, 2016). Each of these is further influenced by environmental factors (Zamai, 2020). This mechanism resembles a type of genetic algorithm through which the generation of new phenotypes intimately links to genome-wide protein–protein interactions that reciprocally interact with viral replication/transcription (Pan et al., 2008). Clearly, there are sets of protein-based biochemical mechanisms that can write on viral and cellular memory by 'intentionally' modifying DNA/RNA, contravening Crick's central dogma.

Perhaps the most consequential reason that the prior presumption of the primacy of genes must be reconsidered is the rapidly increasing understanding of the critical impact of our trillions of co-partnering microbial partners on our biological and evolutionary development (Gilbert et al., 2010; Miller, 2013, 2016b, 2018; Chiu & Gilbert, 2015; Miller et al., 2020a; Torday & Miller, 2020). It is currently estimated that the human microbiome has between 10 and 100 trillion constituents (Thursby & Juge, 2017). Some estimates of the human gut microbiome suggest that there are ten times as many bacterial cells as our personal eukaryotic cells, and their total genetic complement might be 100 times greater than our intrinsic human one. Others dispute that number and insist that the ratio is closer to one to one (Sender & Fuchs, 2016).

However, it bears emphasis that these estimates are for bacterial cells as the most accessible type of microbial constituency for evaluation. None of these estimates include the virome, and nearly every bacterial cell has its constituency of phages (viruses in bacterial cells) that contribute to its metabolic outputs. Furthermore, each of our body sites has its personalized constituent microbiome. For example, human testes have a dedicated microbiome that contributes to testicular function (Altmäe et al., 2019). None of these should be considered passive since microbes function as gene-swapping collectives (Goldenfeld & Woese, 2007).

Our obligate microbiome is a crucial participant in all of our metabolic processes and serves a critical reciprocating role in our adaptive resiliency (Miller, 2016a, 2020b, 2018; Miller et al., 2020a; Torday & Miller, 2020). Over the last several decades, research has confirmed that the microbiome of holobionts has a substantial role in biological development based on extensive mutual dependencies (Kelly et al., 2017). Our constituent microbiome has a considerable impact on our phenotypes. Through that partnering contribution, microbiomes significantly contribute to holobionic phenotypes (Lynch & Hsiao, 2019). Most readers are likely to be familiar with the broad impact our gut microbiome has on our metabolism, partially governing satiety, obesity, and bowel function. However, our microbiome also crucially influences our brain and nervous system. For example, our microbiome is essential for the proper operation of our human gut (Cho & Blaser, 2012), brain, and central nervous system, even affecting our moods and behaviours (Cryan & Dinan, 2012). Further, the microbiome has a critical influence on the entire suite of functions of our immune system, serving as a critical interface between health and disease (Postler & Ghosh, 2017). Research reveals that every aspect of our metabolism and physiology has a co-dependent relationship with our microbiome and adaptive immune responses across our lifespan (Hooper et al., 2012).

Accordingly, it is insufficient to consider evolutionary development through the restrictive lens of a central genome absent the contribution of this essential, obligatory contribution factor comprising a genetic complement of millions of genes that vastly outweighs our paltry 23,000 or so genes. That peripheral, correspondent genome may not be primary, but neither is it inconsequential. What role does this obligate microbial genome serve? It is the ready response system to immediate environmental stresses (Miller et al., 2019). Our relatively insulated central genome reacts more slowly. Together, they comprise our exceptionally competent biological system.

Nor is it realistic to singularly concentrate on DNA. Our vast repertoire of cellular RNA agents represent critical aspects of cell–cell communication, coordination, and regulation (Villarreal & Witzany, 2019). Most viruses, as mostly RNA agents, also participate in this intercellular communication grid. All of these have a complex working relationship with cells, along with TEs, circular DNAs, and RNA stem loops that contribute to every aspect of cellular life, exert regulatory controls, alter adaptive immunity, and participate in evolutionary transitions (Witzany, 2020).

That Was Then, This Is Now: The Demise of the Selfish Gene

The thorough revision of our previous concept of the nature of the genome heralds a complete reappraisal of biological and evolutionary development. In the 20th century, evolutionary thought was dominated by the theory of the 'selfish' gene (Dawkins, 1976). Evolution was the product of natural selection. It was not about species or communities but a narrative of preserving individuals as vessels for their selfish genes. In that era, genes and gene frequencies were thought to rule. Biology in the 21st century is transformative and explicitly cognition based. In 21st-century biology, cellular consciousness is biology's driving force. The central enacting expression of that cellular consciousness is the cell's requisite measurement of the informational cues that the environment contributes. Cells measure because living information is imprecise. Consequently, the cell's sentient capacity to sense that its information is uncertain represents the distinct root of cellular self-awareness, supported by its vast panoply of bioactive molecules, energetic fields, and the full array of its senomic physical apparatus. From that living base, all life spills forward.

To measure better, cells collaborate. This requisite impulse energizes natural cellular engineering and, correspondingly, since cells intimately partner with the virome, natural viral–cellular engineering. The purpose of that engineering is clear-cut. Cellular engineering and, indeed, all cellular communication and the cellular deployment of resources are tasked to solve problems. In cellular terms, problem-solving equates with the maintenance of self-referential homeorhetic balance. Thus, the role of the cell's entire genetic complement, including its central genome and abundant supporting cytoplasmic genetic constituency such as plasmids and circulating small RNAs, clarifies. All the cell's genetic material is part of a flexible read–write informational architecture supporting cellular self-identity as reciprocating, retrievable, and deployable memory purposed for cellular problem-solving. Reproduction must now be viewed differently. Reproduction is another form of problem-solving for cells to sustain the three cellular forms over billions of years. Responsive genomes participate in that process through the self-editing processes of natural genetic engineering and continuous modifications by epigenetic impacts.

If the genome and other intracytoplasmic genetic participants function as memory, what exactly are they remembering? Genomic memory is the repository of the continuous narrative of cellular solutions to environmental stresses. All multicellular organisms are an expression of that narrative in biological form. Through this continuous reciprocating process, cells maintain themselves through constant assimilation of the external environment. To accomplish that, cells form holobionic tissue ecologies that further aggregate to become us, as holobionts. Consequently, evolution can be reappraised beyond a 20th-century narrative of genes to an entirely different, fluid dynamic of intelligent, measuring cells upholding their fates through collaborative information assessment and fluid communication (Miller et al., 2021). Strikingly, biology cannot be a primarily random process. All cellular actions are based on self-generated internal measurements and their further communication to stimulate communal actions. Such deliberative actions represent cellular predictions that are the opposite of random (Miller et al., 2020a, 2020b).

Over 40 years ago, and long before the contemporary research that underpins this chapter, the brilliant physicist Walter Elsasser intuited what still represents the most illuminating concept of the gene (Elsasser, 1981). Working within a framework of quantum physics, he believed that all biological forms fundamentally represent expressions of the superposition of states. The concept of superposition is difficult and non-intuitive but has been confirmed innumerable times. In physical systems, many configurations are possible at any time. Superposition indicates that the most probable state is not any specific fixed one but instead represents a combination of all possibilities that superimpose on each other. Organisms achieve their form by 'settling' into one or another superposition among the many into discrete biological expressions. Elsasser insisted that this type of living system could never be controlled by any simple mathematical rules. Instead, the 'selection' from among this large reservoir of possibilities represents an act of creativity for that organism. Consequently, heredity reproduction is a process of 'creativity with constraints'. Necessarily, these must conform with quantum mechanics, and those constraints are why 'progeny tends to resemble progenitors'. This viewpoint significantly changes the prevailing view that evolution is a mechanistic process that is an endless stream of minor errors, as the modern synthesis stubbornly maintains. Instead, a new biology emerges in which a gene is an 'operative symbol which functions as the releaser of a creative process' (Elsasser, 1981, p. 131).

There is a critical entailment from this edifying insight. Any released cellular creativity derives from the self-referential measuring assessment of information. Although not obvious, any information or cell–cell communication purposed towards creativity as biological expression is necessarily a prediction by both specific cells and collaborating cells as co-engineering. Yet, this coordinated process is precisely how biological variation emerges. Environmental inputs stimulate cells to co-engineer in alternative directions. Hence, biological variations are not random events. Instead, biology variations are purposeful and creative expressions sourced through cellular predictions (Miller et al., 2020a, 2020b).

Additional research is needed to establish the entire role of the genome and its constraints. However, it is now clear that far from being a passive participant, as once believed, the genome is a direct and flexible participant through consistent and responsive self-editing as natural genetic engineering and epigenetic modifications (Witzany, 2011; Shapiro, 2016, 2019). The genetic mobilome, including large numbers of TEs, is crucial for cellular adaptation in cooperation with its flexible genome (Shapiro, 2017). Accordingly, non-selfish genes assist essential symbiotic partnerships among cells and their companion viruses to attain and sustain compatible habitats as niche constructions. Within this narrative, genes are tools. What is the glue that binds these complex processes? Cellular sentience forms that critical bond.

All cells are self-referential cognitive, sentient agents. Computers are logical but share none of these defining attributes. Computers have no ambiguities. They share no doubts, preferences, or experiences, nor can they harness random occurrences, like human engineers often do, in any manner like the living frame. Neither do our genes. As we noted in the Prologue, AI devices have already passed a version of the Turing test and seduced a perceptive human into believing it is fully conscious. Yet, it is still a computer, offering a passing simulation of a breathing human but only remaining an engineering deception.

9

The N-Space Episenome

Life as Information Management

Life Requires Information

The specific point of this volume is an attempt to place the elusive character of consciousness into a robust biomolecular basis. Nonetheless, some attributes of cellular consciousness stand apart from any known biomolecules or membranes. Unquestionably, the elusive sense of consciousness includes some types of non-physical representations that participate in composing our apprehension of reality and thereby govern our lives. As will be explained, this non-physical attribute of the living state is its connection with information space. This chapter will discuss how our relationship with information underpins our existence, how organisms obtain information for internal self-analysis, and how they share information among themselves.

Consciousness/cognition/sentience began some 3.7 billion or more years ago and is coincident with life. We can confidently make this statement since our proof of life is specifically dependent on its embedded collaborative characteristics as a hallmark of cognitive action evinced as microbial mats and stromatolites (Nutman et al., 2019). As single cells do not fossilize, only life in its colonial form leaves evidence of its prior existence. In Chapter 7, we emphasized that this type of colonial life was crucially dependent on the cellular sharing of information. Information must be disseminated and combined among cells since all the information that any cell has is ambiguous, as has been previously detailed (Torday & Miller, 2017). For this specific reason, intelligent cells measure information as an internal self-referential process necessary to assess the its validity (Miller et al., 2020a). Indeed, that sensing of uncertainty and its obligate measurement can be considered the defining characteristic of cellular cognition as central to all cellular problem-solving that is imperative for survival (Miller, 2018).

The Sentient Cell. Arthur S. Reber, František Baluška and William B. Miller Jr., Oxford University Press.
 DOI: 10.1093/oso/9780198873211.003.0009

Further, as previously noted, cells share measurements through abundant communication to improve the validity of their assessment of environmental cues. This drive is elemental and underscores the cellular information cycle introduced in Chapter 7. Importantly, cognitive awareness of information uncertainty as life's condition propels its collaborative assessment.

The foregoing has a remarkable entailment that has received scant attention but is nonetheless absolutely crucial. The cellular process of measuring the validity of environmental information is entirely internal to the cell (Cárdenas-García, 2020; Miller et al., 2020a). Any information that any cell might analyse must traverse an external membrane and then be transmitted multiple times within the internal architecture of any cell across its pertinent senomic apparatus. Consequently, the intracellular analysis (measurement) of information is conditioned within that flux which necessarily imposes substantial noise, distancing any intracellular assessment of information away from its presenting value by imposing potential sources of error (Miller, 2018; Miller et al., 2019, 2020a, 2020b). Accordingly, anything that a cell knows about its environment is based on its exclusive self-referential analysis of that sensory information—and any information that cell has is entirely self-produced. Although the same principle of information insecurity holds true for shared information, collective assessment narrows the range of possibilities beyond the appraisal of any individual cell. All sensed information for cells, even in combination, always has some level of uncertainty and inherent noise due to a loss of integrity through its active transmission from outside the cell to its interior.

Even when cells are self-similar, each is a self-referential agent making its individual measurements. However, the collaborative networking of cells often requires cells of different types to communicate and cooperate. For example, biofilms frequently have constituencies of dissimilar microbial species. All multicellular eukaryotes are complex, seamlessly integrated assemblages of highly differentiated cells and a coexisting and obligatory microbial cohort that at least equals the number of native eukaryotic cells (Miller, 2016b). All must precisely integrate for an organism's survival. Therefore, because multicellular collaboration is one of life's essential requirements, a natural question arises. How might each cell's individual, self-generated cellular measurements be shared within the multicellular whole to yield useful predictions if they are unalike due to differentiation or by being a frankly different species? How might all these disparate separate cellular 'selves' effectively communicate their idiosyncratic measurements of ambiguous environmental inputs to one another to compare them and produce effective biological outcomes?

By default, it must be assumed that collective cellular measurement obliges some common frame of reference to yield consonant and productive biological outputs (Miller et al., 2020b). Cells must have a collective referential measuring platform to collaboratively engineer effective biological products. Consequently, we need to explore the unfamiliar terrain of information space and how cells organize to measure together successfully.

To begin, we need to deal with what ought to seem straightforward but is decidedly not. Exactly what is information? The concept is quite clear to us in our normal, casual conversations. We exchange information all the time. However, among scientists, there has to be a formal definition of information that can be deemed scientifically testable and refutable, and that is precisely where the problems start. Surprisingly, there is no formal definition of information that garners universal acceptance. As a result, how we might define information is equivocal. Some even argue that the term information has no specific meaning at all (Roederer, 2004).

We could write an ocean of words about this topic, and many academics do. However, it is possible to be rigorous and still keep it simple. Information is the sum of the interactions between matter and energy that an observer can appreciate. For most people, an observer is a cognitive agent. Of course, many scientists would argue that information exists even when there is no perceiving cognitive agent. However, for our purposes, information to a living cell is the aggregate of those interactions between matter and energy that represent environmental cues. What matters to living agents are differences in the environment that might upset their dynamic state of balance. Therefore, the cellular meaning of information is a perceptible self-referential difference. Fortunately, one of the most widely accepted formal definitions of information is MacKay's 1969 formulation in which information is deemed 'a distinction that makes a difference', which Bateson (1973) changed to the often quoted, 'the elementary unit of information—is a difference which makes a difference' (Floridi, 2003). Accordingly, in the self-referential living frame, information forms when there is a 'difference that makes a difference' to that self-referential organism, thereby becoming a subject of internal measurement and potential cell–cell communication and bioactive deployment (Miller, 2020a).

Why does a difference matter? Because the interactions of matter and energy, those that constitute external reality, can only be sensed based on a difference sufficient to impact cellular sentience. Only matter–energy differences will stimulate the cell's senomic apparatus to become a subject of internal measurement. The only thing a cell knows is what it can sense.

What Is Information Space-Time?

With that background, let's tackle the unfamiliar concept of information space-time. At first, this type of conceptual frame might seem quite abstract. However, there is good news. All of the complex mathematics that makes physicists giddy is not required. The fundamental concept of space-time is that there is a four-dimensional 'fabric' in which physical objects are embedded and energetic actions occur. Naturally, this concept has been formally expressed in mathematical terms to explain gravity as a curvature of space-time. However, for our purposes, it is easier and more productive to simply accept that there is an observable universe of objects and energy available to interrogation by a self-referential observer as their universal interactions. All of the universe is space-time, and all of that universe is potential information to the appropriate observer.

Accordingly, for any self-referential conscious entity, there is an external universe of potential information that might impact it. For any observer, or any cell for that matter, there is an outward external environment that stimulates its senses. Naturally, that external environment is composed of matter and energy. However, as odd as it might seem, we do not directly sense matter or energy. They only exist through their interaction with one another as it travels across the interstices between cells and crosses the external cellular membrane. What is sensed is the product of biomolecular processing of events and stimuli. For example, on the most simple level, if light does not strike an object to make it classically observable, it is unknown to us if we are limited to experiencing it just through vision. Similarly, every one of our senses requires an interaction between matter and energy, which potentially constitutes information to us. Naturally, it depends on where we are with respect to the stimulus and many other factors. We experience light with an energy spectrum in the 700 nm range as red. But, of course, there is no 'redness' in the light; the self-referential experience emerges from the interaction with particular cellular functions. If you can imagine all of these overlapping cues as a field of potential sources of information, then you would approximate the concept of what information space-time means to any organism. Consequently, all living creatures assess their physical environment through an individual self-referential interpretation of information space-time. Cells experience the environment in exactly the same manner.

Each cell assesses its potential sources of information from a theoretically unlimited field of possibilities from an individual self-referential information space-time matrix which is termed its Pervasive Information Field (PIF)

(Miller, 2016a). Consider this PIF to be the effective boundary of the extent of all possible sources of information available to any specific cell, representing the summary envelope of all the matter–energy interactions that could become a 'difference that makes a difference'. This information field concept was initially conceived by Lloyd (2002) as a means of defining a universal, scale-free, self-organizing system and had been initially applied to concepts of information storage and for describing and modelling social systems (Plikynas et al., 2012). Subsequently, it has been directly applied to biological systems to help explain cellular dynamics (Miller, 2016a, 2018).

This field concept is particularly useful in the context of cells because the transmission of cell–cell signals is more like a broadcast than a targeted communication. Unlike a directed communication network, a field of information directly implies an extensive range of possibilities and inputs without any requirement for exclusivity. Moreover, the concept of fields in biology further implies a potential for overlap, which is easily understood as the intersection of conjoining fields. Within this overlapping matrix, information can be shared and be mutually accessible to a variety of observer-participants and an information field is roughly analogous other familiar fields, such as electromagnetic fields. Just as with these well-understood energy fields, whose effects trail off at a distance, in an information field, information quality from remote sources would be less than those informational cues derived from more proximate sources. In biology, particularly with mediums and membranes, distance from a relevant source is an impactful source of ambiguity, thus helping to understand a cellular context where information is conditioned within uncertainties and noisy distortions (Miller, 2016a, 2018; Torday & Miller, 2017). Harken back to our discussion of communication within and between cellular collectives in Chapter 2 for an example of how this kind of biomechanism operates in prokaryotes.

The PIF concept specifically comports with the self-referential frame as it is an individualized connection to information space defining the limits of 'self'. As the summary field of all the potential informational inputs to any cell, it helps explain the living condition that all the information that any cell has is internally produced. Any stimulus from a cell's information field can only be interpreted after it has passed through the interposed plasma membrane which serves as an integral feature of its self-referential status. Necessarily, that passage imposes conditional uncertainties. Therefore, internally received information must be measured to assess its validity and its meaning vis-a-vis cellular balance. Accordingly, the cellular plasma membrane co-aligns, and indeed, co-evolved with the cell representing the competent interface between the cellular interior and its exterior, governing its sensory inputs (Baluška et al., 2021b). Consequently, all the information that a cell possesses is a function of the senomic apparatus of the cell.

As previously outlined in Chapter 7, the senome is conceived as cell-wide, information architecture, initiating at the plasma membrane as the outer bioactive interface between the external environment of the cell and its complex internal working. Thus, the senome represents a type of sensory organ for assessing environmental cues (Baluška & Miller, 2018). In this way, cellular informational impacts are conveyed across the cell in a highly coordinated fashion, enabling the contingent deployment of cellular resources that ultimately become cellular predictions. The senome represents a pivotal gateway between the internal milieu of a cell and its external environment in its reception of information which, in turn, has its impact on the epigenome and genome, thereby facilitating the continuous assimilation of the environment to maintain crucial self-referential homeorhetic preferences and balance (Baluška & Miller, 2018). From the entire constellation of universal information that a cell might sense, there must be a reliable compact representation that will ultimately permit accurate cellular predictions from its internally self-generated measurements. In this manner, the cellular PIF is an integral part of the cellular senomic apparatus.

The N-Space Episenome Guides Multicellularity

Since each cell has its individual PIF and senome architecture and must necessarily experience the environment individually, how might each of these 'selfish' entities collaborate, cooperate, trade resources, or artfully compete if they are entirely distinct separate and competent selves? In answer, there must be a platform that permits the sharing of information through concordant measurement. Consequently, an N-space Episenome has been proposed that represents just such a space-time platform for the conjoint assessment of information, conceived as an aggregation of all of the individualized cellular PIFs of participants in a multicellular network (Miller et al., 2020b). This collective space-time matrix can be considered a multicellular organism-wide summary informational architecture as a common platform for systematic self-referencing for co-aligned measurements among individual cells. Thus, many types of very different cells can productively collaborate in confrontation with ambiguous environmental cues. These cues derive from external N-space and necessarily impact every individual cell whose cross-communication with other ecological partners permits multicellular cooperation. Just as a PIF represents the field of potential direct and indirect sensory inputs to an individual cell, the N-space Episenome is its multicellular counterpart (Figure 9.1).

The N-space Episenome represents a whole-cell informational field projection. It is not itself directly material, but it achieves functional physicality by its correspondence with biological expression. Information can be deemed physical in cognitive systems since it directly relates to physical degrees of freedom (Walker et al., 2016). Furthermore, there is a direct link between the external environment with its uncertain informational cues and the cellular senomic apparatus. In turn, that cell-wide interface triggers the cell's internal measurement of the incurring information to assess its meaning, which leads to its bioactive deployment as a physical action. Therefore, the N-space Episenome acts as an informational matrix that connects and coordinates the cellular senome, as the summation of the cellular sensory mechanisms, with the genome and epigenome of each individual cell to achieve comporting multicellularity (Baluška & Miller, 2018; Miller et al., 2020b). Consequently,

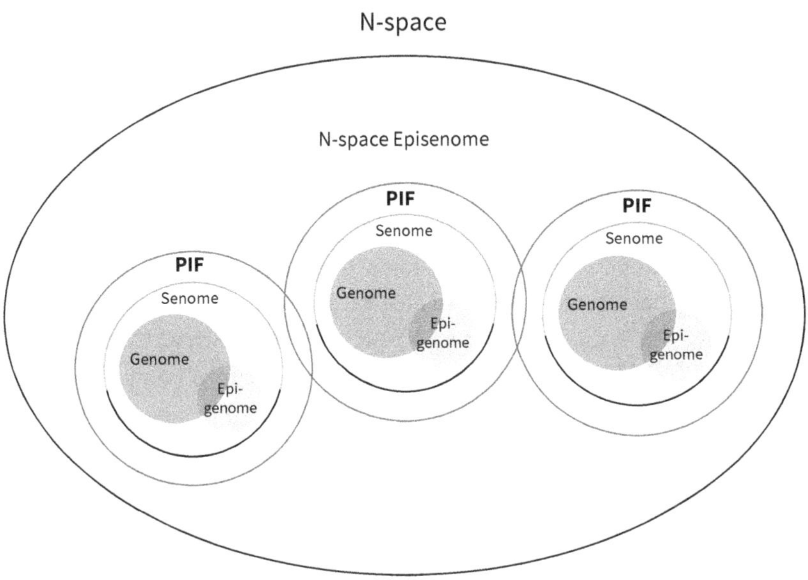

Figure 9.1 The partitioning of N-space. Each cell has a pervasive information field (PIF), which represents the summary field of all potential information inputs that can be received by its senome, genome, and epigenome. In holobionts, these individual cellular PIFs aggregate into a reiterative information field projection as an N-space Episenome. The N-space Episenome information field permits concordant cellular measurement of environmental cues and also serves as a heritable and transferable information architectural matrix templating organismal morphogenesis and development.

the N-space Episenome acts as a reciprocating read–write informational matrix that is the guidepost of multicellular life.

Through this process, individual cells combine their individual measurements of environmental cues based on their individual cellular senomic apparatus and self-appraised internal measurement of that informational meaning. In the aggregate, as a 'wisdom of crowds' writ at the cellular scale, cells improve the quality of their EI* to mitigate environmental ambiguities (Miller, 2018).

It should not be considered controversial to assert that there must be a common referencing platform to permit the productive multicellular work as co-engineering and niche construction that characterizes multicellular life. Although humans and cells do not share the same cognitive frame, both are competent problem-solvers at scale. For us to work together at our scale, we must find and deploy a common means of communication and use common referencing measurement tools for a specific project, such as a 'kilometre', 'metre', or 'dollar'. For cells to work in concert, they must do the same. Certainly, cells share signalling mechanisms to communicate as, for example, in quorum sensing. Since each cell measures separately, by living definition, they must share their senomic measurements of environmental stresses so that they might co-engineer to yield productive biological expressions.

Consequently, an N-space Episenome is a requisite attribute of multicellular life. The appraisal of that necessity is plain enough if we consider our human engineering endeavours. As humans, we readily find a shared means of communication and measurement for any enterprise. Certainly, their absence would render any co-engineering project a chaotic mess. Notably though, we all have a nearly identical species-specific genome and react to stresses using the same human toolbox as our framework. In contrast, eukaryotic multicellular life is radically different. As holobionts, we are vast assemblages of cohabitating differentiated cells and an intimately partnering microbiome. Moreover, it is certain that our constituent microbiome with its trillions of constituents are essential participants in our metabolism, physiological mechanisms, and adaptive capacities (Miller, 2016b). However, there is no doubt that that each of these microbial players is an individual 'self' (Ford, 2009, 2017; Trewavas & Baluška, 2011; Miller et al., 2019; Reber & Baluška, 2022; Reber et al., 2022, Baluška et al., in press). Moreover, each of these microbial partners has a completely different genetic complement compared to eukaryotic cells and uses a diverse toolbox to problem-solve. Yet, all comport with one another to produce harmonious holobionic metabolism, physiology, phenotypes, and cognitive responses to environmental stimuli. Accordingly, an N-space Episenome is a necessity to permit comprehensive, successful, and consistent coordinated responses to both sudden and long-term environmental stresses (Miller et al., 2020b).

How this summary information field is specifically constructed has been directly addressed in detail elsewhere (Miller et al., 2020b; Baluska et al., 2021b). In basic terms, this type of informational architectural field can be productively analogized by its relationship to the non-physical *k*-space used in magnetic resonance imaging. *K*-space is a mathematically generated compact topological space that contains other compact sub-spaces within it. These partitions in *k*-space can be used to generate an anatomical magnetic resonance image as a read–write field projection. Each compact subdivision in *k*-space is its own compartment with its distinguishable data, but also contains a representation of the entire data matrix that has been generated. In this manner, the data make the image, but the image can be used to reverse reconstruct that sub-partitioned data. The inherent reciprocality within *k*-space is similar to how cells share information space as a conjoined N-space Episenome comprised of sub-compartment individual PIFs. Although further details are outside of the scope of this chapter, interested readers are directed to the references.

As a field projection, the N-space Episenome can be regarded as a replete informational landscape and a summary field of all of the conjoined cellular information spaces (PIFs), thus functioning as a shared field of measurement and a developmental canvas with its own dimensions, limitations, and boundaries (Miller et al., 2020b). Accordingly, it is a guiding field of potential shared informational inputs that is in constant reciprocation with biological materiality, impacting its genome, epigenome, and the full range of cytoplasmic constituents, especially its endomembranes. Crucially, all of these are aspects of its senomic architecture enabling real-time cellular responsiveness to environmental cues. Consequently, the N-space Episenome, just as the case for its senome, is intrinsic to cellular consciousness.

Furthermore, it is not sufficient that this shared information platform merely serves as a centrepiece of real-time responses to environmental cues. Research has revealed that every stage of holobionic life has its necessary microbial contribution at all successive stages, including the fetal experience (Miller, 2016b). There is a constituent exposure to microbial influences throughout fetal development, both from maternal microbially supplied metabolites that have transplacental circulation and a small exposure to microbial life *in utero* (Miller, 2016b). Even if these transplacental microbes are few in number in fetuses compared to what is experienced by a newborn, the assault of 'foreign' cells and viruses would be overwhelming if there was not a suitable platform for coordinate responses. Further, embryogenesis requires flawless coordination among a great variety of fully differentiated eukaryotic cells proceeding uninterruptedly and with absolute precision if development is going to be successful. The tiniest migratory errors at early stages would be immensely damaging. Thus, holobionic life mandates a functioning information architecture that must pre-exist birthing. No emergent phenomena could account for this intricately coordinated process (Miller et al., 2020b).

Successive generations of scientists have attempted to identify a templating mechanism for morphogenesis. Although a variety of bioenergetic fields have been proposed that might account for coordinated embryogenesis and further developmental morphogenesis, none of these have been validated (Miller et al., 2020b). All other potential candidates for such a controlling developmental mechanism suffer from a similar deficiency. Each would have to be a product of emergence from cellular multicellularity. Just as with our prior discussion of the 'emergentist's dilemma' for the origin of cognition, the same requirement applies to developmental templating and morphogenesis. Furthermore, cells are not automatons. Each self-generates its individual interpretation of environmental cues. Nonetheless, they must intricately coordinate.

Consequently, the N-space Episenome must be heritable and transferable, functioning as an information-templating architecture and essential information nexus, facilitating the sharing of relevant information from the zygote forward to guide multicellular embryogenesis and holobionic morphogenesis (Miller et al., 2020b; Baluška et al., 2022a). The critical migration of neural crest cells is a good example of that necessity. Neural crest cells represent a transient embryonic cell population that migrates collectively to various locations throughout early embryogenesis, contributing various cell types to several different organs (Szabó & Mayor, 2018). Research has documented that this is an exquisitely coordinated process that proceeds along exact embryonic pathways. Furthermore, neural crest cells exhibit a variety of differential migratory behaviours. At some points along this pathway, these cells co-align in mass-like sheets, edging towards the fetal brain, whereas elsewhere, they migrate in chains towards the fetal trunk. Deploying a range of concordant signals, they manage to collectively move. Yet, research has demonstrated that this migration pattern is not under genetic control (Szabó & Mayor, 2018). Instead, the N-space Episenome acts as a heritable spatiotemporal templating informational matrix, governing cell–cell coordination across multicellular holobionic lifespans.

The further advantage of the N-space Episenome concept is that it offers a framework resolving how epigenetic processes yield coherent organismal responses. Although traditionally overlooked, it is necessarily true that any epigenetic markers are initially expressed at the level of individual cells before they can become organism-wide expressions of phenotypic variation (Miller et al., 2019). How might trillions of individual cells sort epigenetic impacts as valid environmental cues if they must, perforce, be self-interpreted?

It is not only cells that participate in information space. Viruses must have their own form of senomic capacity since they very capably utilize cellular resources and, pertinently, have their own type of nucleocapsid boundary. The senome model is appropriate for viruses since they use a complex arbitrium system to communicate that links to viral memory (Brady et al., 2021). Through this means, viruses link to cellular information space and presumably, then, to their target cellular N-space Episenome, thereby energizing the virocellular symbioses that are the cornerstone of biology. Conversely, cells must reciprocally link to viral information space since they engage in co-engineering for a broad range of problem-solving, and cells must gauge what level of resources can be devoted to any interactions that are volitional (Miller et al., 2021, 2023).

A great deal of environmental information is noise which must be reliably distinguished from more meaningful inputs. Noise is a constant feature in biological systems. Nonetheless, cells do not merely deal with it but can deploy certain forms of noise productively, channelling some either stochastic informational inputs or random events into productive outputs. Noble (2017) termed this biological process the 'harnessing of stochasticity'. However, in order to accomplish this, as Noble and Noble (2018) asserted, '[organisms] must employ a comparator to find the solution that fits the challenge' and to do so, it must have agency and make choices. The N-space Episenome functions in the role of that multicellular comparator as a common reference platform and a form of institutionalized organismal memory that complements retrievable and deployable genetic memory. Thus, the N-space Episenome permits sufficient plasticity to meet current environmental stresses but adheres to a constraining long-term memory repository that represents a counterbalancing constraint against adaptive oversteering (Miller et al., 2020b).

The Relationship of the N-Space Episenome to Consciousness

Although it is common to regard information merely as an abstraction, it is far from that. Indeed, it represents biology's actual propulsive force as it is fundamental to biological expression. Consequently, the flow of information is biology's currency. Since biology begins with conscious self-awareness, that flow must be central to the sentience that defines the living state. Consequently, any understanding of consciousness must be predicated on issues of information management.

To better appreciate that relationship, it is clear that the basic principle we began this chapter with is a critical feature in understanding cognition even in the simplest, unicellular organisms. Information can be viewed as being physical, just like matter or energy, since it represents the interaction between these two elemental universal components, and it is these intersections that actually yield biological outputs. Consequently, the structures in nature can be deemed information (Dodig-Crnkovic, 2022). Accordingly, N-space represents the entirety of the continuous interactions between matter and energy throughout the universe as a universal information matrix. Of course, it is only information if there is an observer to perceive and interpret it. However, all cells are conscious agents, and therefore, all are observers (Torday & Miller, 2016b).

Certainly, many physicists argue that the inanimate is capable of measurement, serving as observers of universal quantum phenomena (Fields et al., 2021). However, we can leave that rather abstruse consideration aside and concentrate on the more readily understandable realm of the living. In that instance, it becomes easier to conceptualize the dynamics of an individual cellular PIF. It can be thought of as the functional boundary of universal information space that is accessible to the cell, even in the most indirect sense. Accordingly, a PIF is a circumscribed compartment of information space that is the complete realm of informational cues that a cell might detect through its senomic apparatus. For convenience, it can be thought of as its 'field of view' concerning otherwise limitless universal informational possibilities. Importantly, this narrowing of a limitless panoply of information is a requisite of conscious self-reference since this circumscription of universal N-space is vital to successful information management in the living state. The reasoning behind this is straightforward. Life is problem-solving and organizing the flow of information to support that is absolutely critical to successful adaptive solutions. The flow of information is a defining parameter of self-referential consciousness since adaptive ability equates with intelligence. Necessarily then, intelligent sorting of any set of a potentially limitless range of informational inputs is essential for information management. Accordingly, biology requires partitioning of any universal information set so that information can be coherently managed. Conscious cells manage information through their constitutive senomic apparatus as their receptive information gateway and initial interpreter of environmental informational cues. Crucially though, the cell derives all of its knowable information that can impact its senome from its

PIF, which is its relevant partitioning of universal information space (Miller, 2016a; Baluška & Miller, 2018).

With this background, the correct biological order clarifies. The senome, with its intelligent plasma membrane, is central to life, and genes are tools of the intelligent cell as vital forms of retrievable and deployable memory as discussed in Chapter 8. It is the cellular senome that actually confronts the environment, receiving bioactive information that stimulates its electrically charged plasma membrane, initiating a cascade of cell-wide responses. Received and initially analysed information is conveyed across the cell to interact with myriad other intracellular participants (cytoskeleton, organelles, intracellular vesicles). In turn, each of these has their individual micro-senomic architectures to ultimately deliver productive intracellular measurements (Baluška et al., 2021b). It is only in this manner that cells can determine whether or not it is worth expending further scant cellular resources through either volitional energy-intensive cell–cell communication or the bioactive deployment of cellular/molecular resources. All of these activities support cellular consciousness. Thus, cellular consciousness must be regarded as a cell-wide organized whole.

As noted above, cellular PIFs combine into a multicellular correspondent mutualizing platform as their relevant N-space Episenome, representing a further unifying compaction of universal information space. That N-space Episenome serves as the critical element of the information management system of the holobiont. Accordingly, just as the cellular PIF is central to cellular conscious self-reference, the multicellular further ordering of information space into a multicellular N-space Episenomic space-time partition must be crucial to its form of aggregate consciousness. Consequently, our human consciousness is a manifestation of multicellular information management that corresponds to a conjoining N-space Episenome. Therefore, just as it has been asserted that our N-space Episenome underlies morphogenetic templating by recapitulating *pari passu* with the critical cellular plasma membrane, so does our capacity for conscious self-reference. Thus, it is no longer surprising that our form of consciousness is species specific even though our genetic makeup compared to chimpanzees is 98.5% identical (Mittleman et al., 2021). Our N-space Episenome is idiosyncratically our own, linked to our species-specific senomic apparatus and distinct attributes of our plasma membrane translating intact from generation to generation, preserving its correspondent connection to information space-time. All are foundational aspects of our consciousness (Miller, 2023). While some might feel that this link between the senomic information processing of individual cells and that of humans is a bit of a stretch, it really is not. As noted in several earlier chapters, evolutionary mechanisms function by building on earlier successful forms and functions. The N-space Episenome was first arrived at by unicellular prokaryotes and has served as an essential platform for those of all species that evolved from them. Put simply: if it works, keep it.

The partitioning of universal N-space into functional partitions that match biological forms is highly explanatory for conscious self-awareness. Unquestionably, consciousness exists. Further, as this book unequivocally asserts, it has existed from the first fully competent cell forward. Still, how conscious self-reference operates remains elusive, just as its origins remain enigmatic. However, that does not mean that no progress has been made. Two indispensable determinants can be readily identified through the information space framework. Conscious self-reference is demonstrably dependent on both boundary conditions and retrievable and deployable memory. In the absence of either, no intelligent life would exist on this planet.

On this planet, biology demands reciprocating liberties and constraints. That exact intersection resides within the still little-explored interstices of our intelligent membranes and the manner in which their integrated functions convey across a living cell with all of its complex, interrelated cellular tools. The result is a summation as a Gestalt, an organized whole. What are those cellular characteristics that are fundamental to conscious self-reference? Sentience requires knowing ambiguity which impels the measurement of uncertain informational cues and the productive management of that information. Thus, the sentient process of information management requires a partitioning of universal information space-time linked to fully competent memory.

These specific requirements for conscious self-reference on our planet carry a significant further implication. The possibility of universal consciousness almost certainly rests upon these same particulars. On Earth, our plasma membrane-based senomic boundary system is exquisitely sensitive to an enormous range of environmental cues. This boundary is lipid dependent and deploys a number of novel mechanisms to sustain the cellular interior, including a variety of gap junctions that discriminatively permit only some environmental molecules across the cellular interface. Further, cells are equipped with an array of retrievable and deployable memory tools permitting coherent adaptation, including a genome that can be faithfully inherited due to robust error correction mechanisms.

For a living organism, information is a difference that makes a difference. We live within the borderline where those differences occur. Every living organism attaches to information space-time and uses bioactive molecules and bioenergetic fields to interpret information. That interpretation is the centrality of our consciousness as the foundation of decision-making.

Consequently, understanding our own consciousness requires the rigorous examination of how individual cells, multicellular organisms, and collections of them critically access, interpret, and deploy information within requisite boundaries.

Thus, life on this planet is the product of two definable linked forces: boundaries and retrievable and deployable memory. Might there be any other intelligent life in the universe? Is there any possibility of pan-consciousness? Those enticing possibilities remain elusive unknowns. However, the requirements of information management that decidedly represent the ground state of intelligence on this planet impose remarkable hurdles. Moreover, since physicists assure that all local processes in the universe are indeed 'universal', then otherworldly intelligence would have to exhibit corresponding attributes of boundaries and memory. Most particularly, a functional partitioning of universal N-space would then be a universal requirement for consciousness. Although still unknown, it is possible that these requisites are very specific to this planet with its exact privileges and limitations. Whatever the case, there is no doubt that earthly cellular consciousness requires a very particular type of boundary system and a broad palette of adjunctive tools that permit a very particular ordered state. As of this moment, there is no indication of any sort of interstellar equivalency.

Conscious intelligence requires a means of ordering information/sensory inputs. Consequently, the conscious cell is competent precisely because it has an informational boundary that represents its limit. Similarly, multicellular companionate life requires its N-space partition as its obligate N-space Episenome. The context of cellular problem-solving mandates this cellular attachment to information space. It is only through its conjoining offices that trillions of eukaryotic body cells and a multitude of co-dependent microbes can produce consonant results generation after generation (Miller, 2016b; Miller et al., 2020a). Necessarily, self-referential consciousness links to effective memory and an intelligent, discriminating membrane. On this planet, without that crucial combination, there would not be any consciousness that we could interrogate.

10
Anaesthetics and Their Cellular Targets

Introduction

The reader might wonder, what is the point of a chapter on anaesthetics in a book devoted to the examination of the foundations of sentience, in particular one based on the principle that life and consciousness are coterminous The point is rather straightforward—as we will see, all species are sensitive to anaesthetics and the question essentially asks itself. Why? Why would even the most primitive unicellular organisms be sensitive to and respond to anaesthetics if they did not feel pain, and did not suffer under stress? As we will see, the biological mechanisms involved in the manner in which cells respond to anaesthetics are not simple and, while still not completely understood, it is clear that considerable biomolecular resources are involved. A core principle of evolutionary biology is that processes that require the expenditure of biomolecular resources are ones that are essential for life. In short, all sentient species respond to anaesthetics simply because they are sentient, have valenced experiences, and feel pain.

Discovery of Anaesthesia via Euphoria and Party Life

Humans used diverse plants and their products for relieving pain and altering consciousness. Many years ago, opium poppy was cultivated by Sumerians beginning as early as 5000 BC (Krikorian, 1975). In a long search for chemicals to relieve pain in humans, a momentous breakthrough was achieved in 1846. On 16 October of this year, surgeon John Warren, in a public demonstration, removed a tumour from patient with help of ether anaesthesia with dentist William T. G. Morton acting as the first anaesthetist. A few months prior, Morton had reported that ether made tooth extraction painless (Bigelow, 1846; Rinaldi, 2014). However, the true pioneer of anaesthesia was a young surgeon, Crawford Williamson Long, who organized social gatherings in 1841 using a novel substance known as 'party gas' where people

The Sentient Cell. Arthur S. Reber, František Baluška and William B. Miller Jr., Oxford University Press.
 DOI: 10.1093/oso/9780198873211.003.0010

inhaled ether in order to attain euphoria. He noted that some confused participants had not recognized either blows or falls and acted as if they had not perceived any pain. Noting this correlation, Long had used ether to remove a small tumour from the neck of a patient, James M. Venable, on 30 March 1842. Unfortunately for Long, there was no official record of this event. There was a serious controversy about the discovery of anaesthesia and even the US Congress was involved in the case known as the 'ether controversy' in 1847 (Boland, 1923; Anaya-Prado & Schadegg-Peña, 2015). Notably, there was even earlier precedence for this type of discovery. Similar parties known as 'laughing gas parties' began in 1799 based on inhaled nitrous oxide, organized by a young chemist Humphry Davy who later became the president of the Royal Society in London. He organized these gas parties in Bristol and noted the pain-killing effects of nitrous oxide inhalation (Hardman, 2017). Unfortunately, these 'laughing gas parties' were not followed by any medical usage of this pain-relieving gas.

Claude Bernard—Sensitivity to Anaesthetics Is Fundamental to Life

The effects of ether on plants was tested and reported almost immediately after its initial surgical use (Clemens, 1848). Claude Bernard (1878) considered susceptibility to anaesthetics a basic feature of life and documented that not only humans but also animals and plants respond similarly to anaesthetics by losing environmental responsiveness and awareness. His famous statement 'all life can be anaesthetized, the rest is dead' dominated the field of anaesthesia in the 19th century (Perouansky, 2012; Grémiaux et al., 2014). His unitary view of life was based on his personal experiments documenting very similar sensitivities and responses to anaesthetics in humans, animals, and plants (Bernard, 1878).

The mystery of anaesthesia includes three fundamental features. First, all life can be rapidly and reversibly anaesthetized; second, specific concentrations of anaesthetics are needed to achieve the anaesthesia; and third, also the recovery from anaesthesia is fast after the removal of the anaesthetic agent. All of these anaesthetic principles were well recognized in Claude Bernard's time but our understanding of these issues remains rudimentary and fragmented (Baluška et al., 2016; Kelz & Mashour, 2019). As we will discuss in the next sections, sensitivity to anaesthetics is an ancient feature which originated and evolved when life was still procellular-unicellular. Obviously, there were specific functions for anaesthesia induced by stress-mediated endogenous anaesthetics relevant for adaptation and evolution of these ancient proto-organisms. As all organisms produce their own endogenous anaesthetics when they are under challenging stressful conditions, it can be proposed that these short-term analgesia and anaesthesia states are important for organismal adaptation and survival (Baluška et al., 2016).

Sensitivity to Anaesthetics Evolved in Unicellular Organisms

Although the research and ideas of Claude Bernard were well known, the myth that anaesthetics act specifically on neurons became a persistent one which ended only after Charles Ernest Overton showed that muscle cells are also sensitive to diverse anaesthetics (Overton, 1901). Moreover, ether and chloroform were shown to suppress fermentation of sugar to alcohol by *Saccharomyces cerevisiae.* This action was considered to be further evidence that living organisms are involved in fermentation even at a time when the doctrine of spontaneous generation still dominated life sciences. Similar to Bernard, also Overton confirmed the sensitivity of plants to anaesthetics (Overton, 1901). We will discuss plant anaesthesia in the next section and Chapter 11. However, the effect of anaesthetics on unicellular organisms is a necessary prelude. Anaesthetics affect diverse unicellular organisms, from minute cyanobacteria up to relatively large protozoa and algae (Baluška et al., 2016; Kelz & Mashour, 2019). In 2008, James M. Sonner proposed that the general capacity to respond to anaesthetics allowed unicellular organisms to cope with diverse stressful situations causing perturbations of their plasma membrane and ion channel activities (Sonner, 2008). Of crucial importance, the first 2 billion of years of life on the planet Earth was unicellular and the evolutionary origins of sentience and cognition are based on prokaryotic and eukaryotic unicellular organisms evolving in the very difficult climatic conditions of the relatively young Earth.

Cellular Targets and the Molecular Nature of Anaesthetic Actions

The mystery of anaesthesia has persisted for almost two centuries despite an immense amount of research devoted to this vital phenomenon. As discussed above, prokaryotic bacteria are sensitive to anaesthetics and, therefore, it is not surprising that mitochondria which evolved from symbiotic bacteria are also sensitive to anaesthetics. As early as the 19th century, Claude Bernard had reported that both cellular respiration and photosynthesis are affected by anaesthetics. After more than 100 years of lipid and protein research, the contemporary science of anaesthesia is still mired in diverse theories which fail to fully explain why and how molecules that vary so widely in their physicochemical properties and sizes can equally cause the same end effect of anaesthesia in all living organisms (Baluška et al., 2016; Kelz & Mashour, 2019). Necessarily, there must be a common denominator that underlies that widespread effectiveness. The only biological substrate sensitive to anaesthetics and linking all these diverse organisms is the excitable plasma membrane which is densely populated with ion channels, transporters, sensors, and receptors, serving as an intelligent bioelectric border between the cellular interior and extracellular exterior.

Charles Ernest Overton and Hans Horst Meyer independently discovered that the potency of anaesthetics is related to their lipid solubility. This correlation between anaesthetic potency and lipid solubility is known as the Meyer–Overton rule of anaesthesia (Perouansky, 2012, 2015). It still holds true but its previously unassailable credibility has been diminished by a mainstream focus that has moved towards proteins as potent hypothetical receptors of general anaesthetics. Although anaesthetics show very different chemical and biophysical properties, the Meyer–Overton rule is valid for most of them and there is no other available theory that comes any closer to explaining anaesthesia better as this largely dismissed one (Sandberg & Miller, 2003). Recently, attempts have been made to reinvigorate the Meyer–Overton rule, updating it through a model focusing anaesthetic action on a bioactive lipid bilayer densely populated by numerous transmembrane and membrane-associated proteins (Sandberg & Miller, 2003; Eger et al., 2008). Even if critical membrane-based proteins should turn out to be the ultimate targets of anaesthetics, lipid-bilayer changes induced by inserting anaesthetics are relevant due to structural changes in the membrane which then alter the conformation of all transmembrane proteins. These reciprocating effects cannot be separated. Indeed, most proteins that are affected by anaesthetics are transmembrane or membrane-associated proteins that populate ion channels as well as diverse receptors and sensors (Urban, 2008; Campagna et al., 2003). Moreover, the hydrophobic pockets of critical proteins, lipid–protein interfaces, and interfacial water arc particularly relevant for anaesthetic action (Kundacina et al., 2016; Riveros-Pereza & Riveros, 2018). Among the numerous anaesthetics, ethylene and xenon are particularly interesting examples as they illuminate the whole mystery of anaesthetics and anaesthesia. Ethylene is an almost forgotten general anaesthetic which is chemically close to ether (Urban & Bleckwenn, 2002; Campagna et al., 2003) and had been commonly used in surgery for over 100 years (Luckhardt & Carter, 1923; Dillard, 1930). Ethylene was a very promising and gentle general anaesthetic with almost no side effects; however, accidents, due to careless handling, led to its being supplanted by other anaesthetics in modern surgery (MacDonald, 1994). Intriguingly, ethylene is known as a plant stress hormone produced in large amounts in ripening fruits (Baluška et al., 2016). In the next sections, we will discuss these important aspects in detail.

Plants are sensitive to both local and general anaesthetics. This fact has been known since 1846 when Clemens reported ether-induced plant anaesthesia using *Mmimosa* leaves and *Berberis* stamens just 1 year after the discovery of ether anaesthesia by William Morton (Clemens, 1848; Bonns, 1918; Eger et al., 2014b). Bonns provided a detailed overview of numerous papers published on the effects of anaesthetics on plants between 1847 and 1918 (Bonns, 1918). Although most of these papers have been forgotten, a fresh initiative is reviving this important field of plant physiology (Pavlovič et al., 2020; Baluška & Yokawa, 2021) despite rather fierce resistance (Draghun et al., 2021; Nick, 2021). Nevertheless, there are convincing reports on using volatile anaesthetics in the immobilization of *Mimosa* leaves at concentrations very close to those used in human surgeries (Okazawa et al., 1993). As we will discuss in more detail in the next section and in Chapter 11, plants represent an important centrepiece within the never-ending debate on cellular targets of the anaesthetics.

In cellular anaesthesia, their organelles, such as mitochondria and chloroplasts, are individual targets of anaesthetics (Sonner & Cantor, 2013; Woods et al., 2021). This is not surprising as both mitochondria and chloroplasts were originally independent organisms which entered into symbiosis with their host cells some 1–2 billion years ago, retaining individual endomembranes. With respect to chloroplasts, the timing of their symbiotic origin is more or less clear (Keeling, 2010; McFadden, 2014), whereas the timing and identity of mitochondrial prokaryotic symbiotic organisms is still debated. One of the consequences of mitochondrial sensitivity to anaesthetics is lowering of ATP levels which have direct impacts on several energy-demanding processes including exocytosis, endocytosis, and cytoplasmic streaming (Ewart, 1903; Jung et al., 2022).

Furthermore, several studies suggest that the cytoskeleton represents a sensitive target of anaesthetics. Both actin filaments and microtubules are reported to be affected by anaesthetics, inhibiting cytoplasmic motility and streaming, as well as the functioning of dendritic spines of neurons (Kaech et al., 1999; Platholi et al., 2014). Not only the cytoskeletal polymers themselves but also their motor complexes, including kinesins and myosins, are reported to be inhibited by anaesthetics (Miyamoto et al., 2000; Yoon et al., 2011).

Pressure Reversal of Anaesthesia Hints at the Physical Basis of Consciousness

One of the most fascinating but seldom discussed issues in contemporary physiology literature is the so-called pressure reversal of anaesthesia, well documented in newts and mice decades ago (Lever et al., 1971). Since then, this fascinating phenomenon has been confirmed in all tested organisms (Vadakkan, 2015; Weinrich & Worcester, 2018), strongly suggesting that there are some fundamental biophysical features underlying anaesthesia. A similar response to pressure has been validated in plants, where pressure reverses promotion of seed germination via anaesthetics (Hendricks & Taylorson, 1979, 1980). All this suggests that spatial distances among some critical biomolecules are relevant for both the onset of and recovery from anaesthesia in all organisms. This spatial aspect of anaesthetics and their impacts on cells and their ordered biomolecules can be explained by either binding to critical hydrophobic pockets of relevant proteins or by their insertion into biological membranes, altering their biophysical properties and excitability. As we will discuss below, this lipid ordering and membrane structure have been reported as relevant features of anaesthetic action, further supporting lipid-based theories of anaesthesia. Unfortunately, some reports failed to confirm this phenomenon and it remains a relatively obscure observation. Additionally though, pressure alone can immobilize animals and then, surprisingly, anaesthetics induce recovery of their pressure-induced immobility, known as an 'awakening effect' of anaesthetics (George & Pandit, 2021).

Ordered Lipids and Lipid Rafts as Subcellular Targets of Anaesthetics

Ever since Meyer and Overton independently discovered a strong correlation between anaesthetic potency and solubility in oil and codified it as the Meyer–Overton rule, the lipid theory of anaesthesia has become generally accepted. From 1900 through the 1980s, lipid theory was the dominant one, but it lost its primacy after the discovery that ion channels and several ligand-gated sensors at the plasma membrane also represent targets of anaesthetics (Campagna et al., 2003; Urban, 2008). Currently, the dominant view is that there are multiple target sites in cells for the anaesthetics, and this seems to be one of the reasons that our understanding is progressing rather slowly.

One of the recent discoveries strongly supporting a lipid-based theory of anaesthesia are so-called lipid nanodomains (rafts) representing highly ordered domains of structural lipids (Lingwood & Simons, 2010; Simons & Sampaio, 2011). These lipid rafts represent a silent revolution in our understanding of biological membranes basing a new perspective based on lipid homogeneities in concurrence with a well-validated fluid mosaic model (Singer & Nicholson, 1972). The fluid mosaic model conceptualizes the plasma membrane as a mosaic of a large palette of bioactive molecules including phospholipids, cholesterol, proteins, and carbohydrates, giving the membrane a fluid character. Initially, the concept of lipid rafts met strong opposition (Edidin, 2003; Munro, 2003) but they are now accepted and Nicolson updated the fluid mosaic model to include lipid rafts in 2013 (Nicolson, 2014; Nicolson & Ferreira de Mattos, 2022). Importantly, lipid rafts are organized via evolutionarily ancient proteins containing prohibitin-likedomains (PHB) including the Reggie/Flotillin family.

These proteins are present in membranes including in bacteria, protozoa, plants, and animals (Morrow & Parton, 2003; Rivera-Milla et al., 2006). They organize lipid nano-domains (rafts) which then provide platforms for diverse signalling proteins involved in assembly of endosomal pathways (Langhorst et al., 2006; Ma et al., 2022) serving as entry points for viruses including the SARS-CoV2 virus (Sorice et al., 2021; Bakillah et al., 2022). Lipid rafts based on sterol-derived, ordered structural lipids provide a unique microenvironment for specific sensors and other signalling biomolecules representing targets for anaesthetics (Weinrich & Worcester, 2013; Booker & Sum, 2013). Interestingly, the pressure which reverses anaesthetics also promotes lipid ordering and formation of domains in model lipid raft membranes (Worcester & Weinrich, 2015).

Anaesthetics and Electrons: Hydrophobic Pockets of Proteins Sensitive to Anaesthetics

Proteins with hydrophobic pockets are also relevant to the pressure reversal of anaesthesia (Eckenhoff, 2008; Urban, 2008) in which weak van der Waals–London forces are involved in electron donor/acceptor events, mediating the sensitivity of these proteins to anaesthetics (Vos et al., 1993; Hameroff, 2006). In particular, general anaesthetics inhibit electron mobility within these hydrophobic pockets of sensitive proteins which act as targets of general anaesthetics (Johansson et al., 2003; Hameroff, 2006). Importantly, mobile electrons participate in anaesthetic actions. For example, the respiratory mitochondrial complex I based on mobile electrons is inhibited by anaesthetics in single-cell organisms, plants, and animals including humans (Kelz & Mashour, 2019; Jung et al., 2022).

A new twist in understanding anaesthesia was provided by Luca Turin and co-authors reporting that there is a connection between electron currents in cells and general anaesthesia (Turin et al., 2014; Turin & Skoulakis, 2018). Recently, Smith et al. (2021) reported that radical pairs of electrons are similarly important in this respect. Intriguingly, xenon behaves like all other anaesthetics in this respect, even when it is chemically inert and physically shapeless (Turin et al., 2014). Electron mobility might also be behind the so-called electro-anaesthesia phenomena, when pulsed electrical currents have been reported to induce anaesthesia-like states with reversible loss of consciousness (Papir-Kricheli & Magnes, 1982; Francis & Dingley, 2015).

Importantly, the proteins sensitive to anaesthetics are so-called electronic proteins and are equipped with hydrophobic pockets that use mobile electrons to accomplish dynamic changes in three-dimensional conformations. Anaesthetics bind and fill resulting hydrophobic pockets, causing inactivity of these proteins and freezing their conformation (LaBella et al., 1999). This protein effect may represent one of major impacts of anaesthetics on cells and their life-supporting activities beyond their effects on the bio-membranes. As many of these proteins are localized within membranes and transmembrane sensors, receptors, and ion channels, they are relevant for the dynamic assembly and maintenance of senomic fields (and were discussed in Chapters 6 and 9). Consequently, it is highly probable that the senome and its cellular and supracellular fields are also affected by anaesthetics.

Ancient Origins of Action Potentials Sensitive to Anaesthetics in Plants and Animals

In all multicellular organisms, just as in unicellular ones, the most obvious impact of anaesthetics is sudden cessation of mobility (Kelz & Mashour, 2019). In both animals and plants, electric action potentials animate movements and these stop with immediate exposure to anaesthetics with impacts on both critical transmembrane proteins and the lipid-bilayer of membranes. With action potentials, there is rapid and brief reversal of membrane voltage polarity (Andersen et al., 2009) directly impacting vesicle behaviour at the plasma membrane (Baluška & Wan, 2012). Moreover, sensitivity of the actin-based cytoskeleton is also relevant insofar as tissue and organ movements in plants and animals are based on actin-myosin interactions. Importantly, these action potentials apparently evolved from very ancient plasma membrane repair processes, most probably originating in unicellular organisms (Goldsworthy, 1983; Brunet & Arendt, 2016). Pertinently, this plasma membrane repair system resembles the neuronal synaptic transmission apparatus (Steinhardt et al., 1994). And notably, the cellular events associated with long-distance action potentials link to, and originate from, ancient cellular stress responses corresponding to the close linkages of anaesthetics with cellular stress responses at the bioelectric cell periphery (Baluška & Wan, 2012; Baluška & Mancuso, 2019). All this is important for our understanding of why anaesthesia is central to our understanding of the life. The Claude Bernard unitary hypothesis of anaesthesia is getting strong support.

Endogenous Anaesthetics for Coping With Stress

Dual physiological and evolutionary roles for anaesthesia would be logical from the perspective of the unitary hypothesis of anaesthesia, asserting this phenomenon as the most obvious definition of life. As Bernard maintained in 1878, it is ubiquitously necessary for coping with severe stress at cellular and higher levels of biological organization (Selye, 1975). Numerous studies have reported that heavy stress induces the endogenous production of substances having anaesthetic and analgesic effects on humans and diverse organisms including prokaryotic bacteria, unicellular protists, fungi, plants, and animals (Baluška et al., 2016). Hans Selye postulated a unified theory of stress as part of a mysterious adaptation energy which allows effective repair and adaptation processes (Selye, 1975; Coleman, 2012). In humans, anticipation of pain induces the endogenous biosynthesis of opiates and cannabinoids both of which lower pain perception (Vaughan, 2006; Corcoran et al., 2015). This adaptive phenomenon is well known as stress-induced analgesia (Ferdousi & Finn, 2018; al'Absi et al., 2021). It allows humans and animals to survive extremely stressful insults, including devastating wounds and presumably evolved very early. Plants are also known to produce numerous endogenous substances which have anaesthetic and analgesic effects (Baluška et al., 2016). These painkillers of plant origin were discovered by humans very early in our history with the first documented examples of their use going back more than 5000 years. Another example of the mind-altering psychoactive substances in plants are indole alkaloids (Omar et al., 2021; Wibowo et al., 2021); including the *N,N*-dimethyltryptamine (DMT) which is found to be produced not only in plants, but also in animals including humans (Barker, 2018; Cameron & Olson, 2018).

As James Sonner and Robert Cantor (2013) concluded, consciousness and endogenous anaesthetics obviously evolved together long before the evolution of nervous systems, beginning with the advent of cellular life several billion years ago. They highlighted ethanol as the best evidence for the importance of endogenous anaesthetics and anaesthesia in biological evolution. As ethanol is toxic to all organisms, one would expect that a long evolutionary period of spontaneous mutations would eliminate responses to ethanol. However, as that is not the case, it can be surmised that the mutations that would be required to block the toxic impact of ethanol on organisms would be more harmful to the organism than the ethanol production itself. Correspondingly, some beneficial effects of ethanol might also be present in smaller amounts. Plants produce endogenous ethanol under stress along with other useful anaesthetics (Baluška et al., 2016; Tsuchiya, 2017). Furthermore, impacts of ethanol on plant root apices (Kagenishi et al., 2022) resemble the effects of other anaesthetics on root apex cells (Yokawa et al., 2019).

Syncope, Thanatosis, and Interoceptive Paralysis of Cells

Sudden loss of consciousness is behind the phenomenon known as thanatosis or feigning of death (Rogers & Simpson, 2014; Humphreys & Ruxton, 2018) and appears to be connected to an overload of interoceptive sensory systems due to an extreme danger. It has also proved to be a successful strategy to survive under diverse life-threatening situations (Skelhorn, 2018; Asakura et al., 2021). Closely related to thanatosis is syncope which is the sudden loss of consciousness associated with loss of postural tone (Alboni & Alboni, 2014; da Silva, 2014). Thanatosis is highly conserved and an ancient anti-predatory survival strategy in animals. There are some common issues connecting the thanatosis with near-death experiences recorded in humans (Kondziella, 2020; Peinkhofer et al., 2021). Importantly, unicellular organisms, such as budding yeast, respond to environmental stress with a sudden loss of metabolism and intracellular shutdown. In addition, the simple perturbation of the plasma membrane with local anaesthetics is enough to induce this paralysis in unicellular organisms (Uesono, 2009; Uesono et al., 2008, 2016). These findings suggest that endogenous anaesthetics could also be behind life-protecting phenomena such as syncope, thanatosis, and other death-feigning phenomena.

Summary

This exegesis on the biomolecular manner of action of anaesthetics clearly supports our CBC model. As noted, one of the keys to our approach is that functions, mechanisms, and processes that were adaptive, with significant Darwinian fitness, are not lost. They are maintained and become the foundation for evolutionary change over time. Showing that all species, all organisms are sensitive to anaesthetics and that all share underlying biomolecular modes of operation reinforces the proposition that all species, all organisms, feel, experience pain, and react to stress—including the simplest prokaryotes, bacteria, and archaea. In short, all organisms have valenced experiences—they are sentient. In Chapter 11, we will take a similarly deep look at sentience in plants where the biomechanisms differ but the underlying proposition that all living organisms are sentient holds.

11
Plant Sentience

Linking Cellular Consciousness to the Cognitive and Behavioural Features of Plant Life

Ancient Origins of Plants in Algal Protists: Similarities With Photosynthetic Animals

Plants diverged from animals at the unicellular level by the endosymbiotic capture of ancient cyanobacteria which evolved into their chloroplasts some 1.5–2 billion years ago. Hence, plants and animals share approximately two-thirds of their 1.5 billion years of evolution and, of course, have common biological roots (Bouteau et al., 2021; Gilroy & Trewavas, 2022). The reason why plants do not have brains or any devoted sensory organs directly links to their sessile nature. Being unable to escape from herbivores to become the basis of the land food chain would not have been possible if they had not evolved rooted bodies (Trewavas, 2014; Gilroy & Trewavas, 2022). A reasonable speculation is that the ultimate reason for the sessile nature of plants was the acquisition of photosynthetic bacteria and switching to autotrophic nutrition using these symbiotic organelles. Nevertheless, it is clear that when animals obtained functional photosynthetic chloroplasts and switched to autotrophy, they also acquired a sessile lifestyle as also exhibited by species such as corals (Baluška & Mancuso, 2009a, 2009b). Additional support for this perspective is provided by sacoglossan sea slugs. These slugs use 'stolen' chloroplasts for their nutrition (Rumpho et al., 2000). This incorporation and maintenance of algal chloroplasts in animal cells is termed kleptoplasty and these chloroplasts are capable of being maintained within cells of the animal digestive system for extended periods of time (Händeler et al., 2009; Pierce & Curtis, 2012). Kleptoplasty allows sea slugs to survive for as long as a year without any organic food from their environment (Laetz et al., 2017; Cartaxana et al., 2021).

The case of corals is interesting in that their long-term symbiosis with photosynthetic algae is possible only because they are sessile organisms. Although corals are animals, they harbour single-celled photosynthetic dinoflagellates and are characterized by a sessile lifestyle allowing the assembly of ecologically

The Sentient Cell. Arthur S. Reber, František Baluška and William B. Miller Jr., Oxford University Press.
 DOI: 10.1093/oso/9780198873211.003.0011

complex coral reefs (Wood, 2003; Blackall et al., 2015). Corals also show several other plant-like features, including modular metameric bodies and a high capacity for regeneration (Davy et al., 2012; Rosset et al., 2021). The symbiotic integration in corals, however, is less stable than in plants. Algal symbionts leave coral host cells under stress, a phenomenon that accounts for coral bleaching. Currently, coral reefs are facing a global crisis due to warming of the oceans and other less well-understood climatic shifts (Bhagooli & Hidaka, 2004; Weis, 2019). However, corals and their reefs are known to recover their symbiotic partners, enabling reef rejuvenation after mass extinction events (van Oppen & Medina, 2020) since coral-based symbiotic networks are robust, demonstrating high recovery potentials (Boilard et al., 2020; Williams & Patterson, 2020).

Interestingly, besides corals, there are several other examples of animals capturing chloroplasts and using them for their autotrophic nutrition including hydra, sea anemones, sponges, and ciliates such as *Paramecium* (Venn et al., 2008; Miyokawa et al., 2022). The common theme in all of these animals is that many of the supporting features resemble plants. In other words, these animals evolved several plant-specific attributes due to their symbiotic chloroplasts and photosynthesis. This is relevant as it reveals that plants are not as different from animals as is generally believed (Baluška & Mancuso, 2009a; Bouteau et al., 2021). Accordingly, there is a tight co-evolutionary path among pollinators, frugivores, and seed disperser partners—including us humans.

From Chlamydomonas to Cognitive Stress-Induced Multicellularity

One of the best understood unicellular green algae is *Chlamydomonas reinhardtii* which has served as a cytological model for decades. Its full genome was acquired and published in 2007 (Merchant et al., 2007; Blaby et al., 2014). *Chlamydomonas* represents an excellent model system to study the evolution of multicellularity. Under stressful situations, it switches to multicellular assemblies (Ratsliff et al., 2013; de Carpentier et al., 2022). One of the strategic advances of multicellularity is that a larger body size is an advantage against predators since it no longer fits within the primary predator's prey scale. Indeed, Herron and colleagues reported that when *C. reinhardtii* is exposed to its ciliate predator *Paramecium tetraurelia,* they evolve to multicellularity, proving an effective defence against predation (Boyd et al., 2018; Herron et al., 2019). Another natural predator, *Peranema trichophorum* also induces transient aggregation of *Chlamydomonas* which reverts back to its unicellular state when these predators are removed (Sathe & Durand, 2016). It can be expected that the evolution of multicellular colonies that formed ancient unicellular algae (Herron et al., 2009; Herron, 2016) was also accomplished under pressure from their predators. Joining of forces, the coordination of cellular activities (Kirk, 2005; Umen & Herron, 2021), and associated sociality (de Carpentier et al., 2019, 2022) are cognitive achievements, guided by cellular cognition, sentience, and communication. Similar examples of stress-induced multicellularity are known among bacteria (Claessen et al., 2014; La Fortezza & Velicer, 2021) and social amoebae (Bonner, 2009; Schaap, 2021). It is possible that individual cells/organisms make cognitive decisions to join forces after evaluating their environmental landscape as dangerous for their survival (La Fortezza & Velicer, 2021; Schaap, 2021). In this new evolutionary perspective, the origin of multicellularity is a cognitive phenomenon based on cellular sentience.

Dinoflagellate Algae as Hunters Enjoying Protist Vision

At the cellular level, organisms harbouring photosynthetic chloroplast in their cells—including corals, algae, and plants—are more complex than fungi and animals. For example, *Paramecium bursaria* harbour lower mitochondrial numbers after hosting endosymbiotically green algae *Chlorella variabilis* (Kodama & Fujishima, 2022). Importantly, endocellular symbiotic organelles have been found to organize camera-like sensory ocelloids allowing a type of vision in protists (Hayakawa et al., 2015; Gavelis et al., 2015). In this newly discovered protist eye, symbiotic mitochondria act as cornea-like structures and modified chloroplasts perform retina-like functions (Gavelis et al., 2015; Yamashita & Baluška, 2023). Interestingly, chloroplast thylakoids also act as retina-like organs in the eyespots of *Chlamydomonas* algae (Kreimer, 2009; Ueki et al., 2016). Vision used by algal hunters is strong evidence for a kind of strategic information management and problem-solving cognitive abilities based on cellular sentience. In the next section, we will discuss similar cognitive phenomena in sex cells (gametes) during sexual reproduction in algae, plants, and animals.

Cognition of Protist-Like Sex Cells (Gametes)—The Special Case of Flowering Plants

In recent conceptual papers, we have discussed sentience and cognition in unicellular protists and asserted an analogical framework in protist-like sex cells (gametes) of multicellular organisms (Baluška et al., 2022a, 2022b, in press). Whereas oocytes (in plants termed egg cells) resemble rather slow, actin cytoskeleton-based amoeba-like cells, sperm cells very closely resemble flagellated protists propelled via cytoskeleton-based microtubules. Intriguingly, this is not a new observation. Nearly 150 years ago, Ernst Haeckel noted that oocytes of *Olynthus* sponges have amoeba-like features (Haeckel, 1878) and some *Porifera* oocytes even feed on bacteria in a fashion similar to contemporary amoebae (Sciscioli et al., 1994; Lanna & Klautau, 2010). On the other hand, sperm cells move rapidly using their eukaryotic flagella propelled via a microtubular cytoskeleton, thereby closely resembling flagellated protists (Avidor-Reiss et al., 2019, 2022). The central part of the sperm flagella regulatory complex is a centrosome-centriole apparatus (Schatten et al., 2011; Khanal et al., 2021). Although flowering plant sperm cells lack flagella, this is a secondary loss as most non-flowering plants have flagellated, motile sperm cells (Renzaglia & Garbary, 2001; Norstog et al., 2004). Despite their rather large evolutionary distance, there are further common issues connecting sperm cell biology and behaviour in plants and animals, especially their nuclear migrations and fusions (Kawashima et al., 2014; Fatema et al., 2019).

In all organisms, sperm cells are highly motile and all their activities are focused on their prime task of finding receptive oocytes/egg cells for cell–cell fusion. In less complex animals such as sponges, in some instances sperm cells are released into water by one sponge and taken up by another for sexual reproduction. In this case, once sperm cells are acquired, they are first internalized into special cells known as choanocytes which transport them through internal tissues up to the oocytes. Other sponges release both sperm cells and oocytes into the water and the sperm cells search for freely floating oocytes for potential fusion. The first mode is closer to the situation in animals and plants when sperm cells are introduced into female bodies and then actively search for oocytes. As this is a difficult task, it is not surprising that sperm cells rely on cellular cognition based on senomic sentience to fulfil their task. Consequently, it is also not surprising that animal and human sperm cells show many neuronal features and express various neuronal proteins (Zitranski et al., 2010; Matos et al., 2021). There is extensive communication between the cells of the female human reproductive tract and navigating sperm cells which is essential for the success of the sexual reproduction (Yeste et al., 2017; Zaferani et al., 2021). From some 100 million sperm cells deposited in the female tract in humans, only a small fraction will reach the oviduct (Lefièvre et al., 2009; Reynaud et al., 2015) in which the oocyte is waiting for competent sperm cells. Undoubtedly, this is a complex, difficult task for sperm cells. Hence, both cognitive sex cells and large numbers of them are required to ensure successful fertilization.

Nitric oxide released by cells of the female reproductive tract assists sperm cells in their navigation (Machado-Oliveira et al., 2008; Lefièvre et al., 2009). Intriguingly, pollen tubes of flowering plants that carry plant sperm cells also use nitric oxide as a signalling molecule for cellular navigation through their style (stalk of a pistil) tissues to reach plant oocytes (Prado et al., 2011). In addition, navigating the pollen tube towards plant egg cells also relies on GABA gradients and glutamate as signalling molecules, activating plant-specific glutamate receptors (Palanivelu et al., 2003; Michard et al., 2011). Non-flowering plants have freely swimming flagellate sperm cells with the exceptions of cycads and gingko trees that keep flagellated sperm cells entrapped within pollen tubes (Klink & Wolniak, 2001; Vaughn & Renzaglia, 2006). Intriguingly, pollen tubes are tip-growing cells, resembling neuronal axons (Lev-Yadun, 2001; Palanivelu & Preuss, 2001), which navigate towards egg cells via cognitive pathfinding as these are deeply buried within stigma (a specially adapted portion of the pistil modified for the reception of pollen) tissues (Kessler & Grossniklaus, 2011; Palanivelu & Tsukamoto, 2011). This close similarity between the intrusive and polar growth of neuronal axons and pollen tubes, guided via similar neuronal principles, are excellent illustrations of the cognitive aspects of plant sexual reproduction and the higher levels of complexity of flowering plants. Although pollen tubes lack flagellate, motile sperm cells, their specialized cells are rich cognitive players in complex plant sexual reproduction.

Plant Cognition and Behaviour—From Root Foraging, via Manipulation of Pollinators and Bodyguards, up to Plant Mimicry

Plants live a sessile life which requires exquisite sensory systems to experience and properly evaluate a challenging environment and adapt effectively. Invasive root systems are essential for plant nutrition and anchoring in soil. These complex systems use root-specific sentience to search the soil for water and those mineral elements that are essential for plant nutrition. Root apices are endowed with plant-specific intelligence and cognitive capacity to achieve this difficult mission in a dark, challenging underground environment (Baluška & Mancuso, 2021; Baluška et al., 2021b). Intriguingly, a similar strategy is used by parasitic plants. These plants will extend a haustorium, a slender projection, from their roots, enabling the parasite to penetrate its target plant tissues and absorb nutrients from it (Yoshida et al., 2016). The only difference is that the parasitic plants do not bother to search for water and minerals in a dangerous underground soil environment. Instead, they tap directly into the relevant nutritional streams in host plants. Intriguingly, they often use sophisticated sensory systems to find the most promising host plants. For example, dodder (*Cuscuta*) uses plant-specific versions of sniffing (Runyon et al., 2006; Mescher et al., 2006), perhaps comparable to a form of vision (Parise et al., 2021), to identify the most suitable host plants. Additionally, there are recent reports on plant-specific hearing where sound waves strong enough to activate relevant plasma membrane receptors inform plants about important proximal environmental cues, including streaming water in the case of roots or buzzing pollinator insects in the case of flowers (Rodrigo-Moreno et al., 2017; Veits et al., 2019).

In leaves and stems, trichomes (fine outgrowths or appendages) act as sensitive plant hearing devices (Liu et al., 2017; Peng et al., 2022). Even further, there are other more exotic and surprising sensory faculties of plants. For example, some plants are capable of sensing electric fields generated by buzzing insect pollinators such as bumblebees (Clarke et al., 2017; Vallejo-Marín, 2019). Gagliano et al. (2012) suggested that some plants deploy root apex sonar-like root navigation and the growing maize root apex is known to generate regular clicking sounds which could inform the growing root apex about nearby physical properties, especially the presence or absence of water in the soil ahead of its projected growth trajectory (Baluška et al., 2021b). Relevantly, maize root apices generate abundant electric spikes (Masi et al., 2009) which might be critical for their cognitive faculties.

Foraging plant roots assess diverse signals of both abiotic and biotic origin to localize and find water enriched with critical mineral nutrients. Roots of the same plant team-up together, showing signs of swarm intelligence to fight, deter, and kill roots of competing neighbour plants (Novoplansky, 2009; Shemesh et al., 2010). Roots use specific exudates to define their identities and monitor their neighbour plants (Gruntman & Novoplansky, 2004; Biedrzycki et al., 2010). The relevance of these root processes permitting kin and self/non-self recognition have been investigated and are being practically utilized in plant agriculture (Yang et al., 2018; Anten & Chen, 2021). However, until recently, the cellular background of these intelligent plant behaviours had remained obscure until the CBC model of cellular sentience provided a novel, theoretical platform for understanding the relevant scientific questions and helping to guide experimental studies.

Although much is being learned about the cognitive abilities of flowering plants, there is still much that remains unknown. For example, the factors that enable recognition of host plants by parasitical climbing vines have not yet been completely clarified. Nor is it fully understood how this action can sometimes be associated with mimicking of these host plants nor the parasitical plants' abilities for recognizing, attracting, and manipulating their host's potential pollinators.

As plants are sessile, they need some assistance to spread their male sex cells which can be enclosed within pollen grains. The most readily available mechanism for that dispersion is relying on the wind for pollination. However, some flowering plants have entered into tight co-evolution with their pollinators which act as assistants in their sexual life. In order to attract these pollinators which are typically insects, they lure them with attractive colours, odours, and shapes and also provide them with sugar-rich liquid nectar. Nectar secretions increase in those flowers that are often visited by their pollinators. Besides sugars, nectars are enriched by other often addictive substances, including caffeine (Nepi et al., 2018; Wink, 2018; Mustard, 2020). These biomechanisms allow plants to control the behaviour of their pollinators and select which of them are most welcome to provide this service to them (Kessler et al., 2010). Intriguingly, GABA and diverse other neuroactive substances increase the cognitive abilities of bees and other pollinators (Nepi, 2014; Carlesso et al., 2021). Besides the flower nectar, plants also produce extra-flower nectar to manipulate insects to provide further services for them. An estimated 8,000 plants use extrafloral nectaries to manipulate insects (Grasso et al., 2015). For example, acacia trees use extra-floral nectaries to attract and manipulate ants, increasing their aggressivity, thereby providing plants some protection from diverse herbivores (Grasso et al., 2015). Ants associated with acacia trees show extremely aggressive behaviour. They will attack animals, even elephants and antelopes, and certainly deserve their insect bodyguard status (Bentley, 1977; Grasso et al., 2015).

A further conspicuous example of the dominant role of plants over their insect 'servants' is provided by orchids which attract their insect pollinator only by providing abstractive clues such as shapes, colours, and forms mimicking female partners (Schiestl, 2005; Jersáková et al., 2006). This mimicry is convincing enough that the fooled males will try to repeatedly copulate with their female-mimicking flowers (Gaskett et al., 2008). A further baffling example of plant mimicry is provided by the Amazonian rainforest woody vine *Boquila trifoliolata* which is able to mimic several host plants with differing leaf shapes (Gianoli & Carrasco-Urra, 2014; Gianoli et al., 2021). This plant mimics different aspects of their host plant leaves, including shapes, sizes, textures, and colours. It remains a mystery how *Boqguila* is capable of this level of complex mimicry (Baluška & Mancuso, 2016; Gianoli, 2017).

Surprisingly, a recent experiment with plastic artificial host plants suggests that some kind of plant vision might be involved (White & Yamashita, 2022; Yamashita & Baluška, 2023). This effect, if replicated, would explain why *Boqguila* can mimic such diverse physical features as shape, size, and colours of host plant leaves. Although generally ignored but especially surprising, plant mimicry is demonstrated by weeds grown with domesticated crop plants, a phenomenon known as 'Vavilovian mimicry', named after Nikolai Vavilov who dubbed them 'secondary crops' (McElroy, 2014). This effect has particular agricultural significance since it is thought that oat and rye plants evolved from weeds into important crop plants.

Almost all current literature on plants delves into evolutionary theories based on genetics and gene population frequencies (Huang et al., 2018; Ye et al., 2019). However, and of vital importance, plant sensory systems and cognition are largely ignored. Undoubtedly, however, it would be important to devote substantial attention to plant sensory systems, cognition, and sentience (Baluška & Mancuso, 2021; Yamashita & Baluška, 2023) to gain any valid understanding of evolutionary mechanisms.

Root-—Fungal Networks: Supracellular Sentience Is Relevant for the Gaia Concept

About 700 million years ago, ancient, small, and simple plants moved from water to land together with ancient fungi. This first colonization of the terrestrial habitat represents the decisive event in evolution of complex life on our planet (Becker, 2013; Žárský et al., 2022) beyond the crucial endosymbiotic fusion of Prokaryota to yield Eukaryota. With respect to the water-to-land transition, the favoured hypothesis is that the first land plants somehow evolved from streptophyte algae. However, ancient algae alone could not have launched this transition since they have no physiological features that would permit them to transform the barren stone land into a life-supporting habitat with lasting niche constructions. In contrast to these algae, ancient fungi were able to assist that transition through the weathering of rocks, thus enabling the tight symbioses between plants and their symbiotic fungi. One can imagine that this interaction might have begun first as a loose association, just as can be observed with modern lichens, and only later energized the algae and fungi mergers that generated the first land plants (Jorgensen, 1993). In fact, most data strongly suggest that the first land plants needed very tight interactions with their symbiotic fungi (Remy et al., 1994; Wang & Huang, 2021). In addition to their role in rock weathering, fungi also assisted the early land plants in coping with high levels of light stress and dramatic temperature shifts, as well as with their problems finding water and mineral nutrients on dry land (Pirozynski & Malloch, 1975; Selosse & Le Tacon, 1998). Although fungal hyphae networks are well studied with respect to uptake and transport of water and mineral nutrients, published data reveal that they transmit diverse signals and information-loaded molecules (Simard, 2018; Fukasawa et al., 2020). These networks are also active in bioelectric communication supported by linked action potentials (Volkov & Shtessel, 2020; Thomas & Cooper, 2022). All these new discoveries about reciprocating interactions are consistent with the Gaia concept (Markoš, 1995; Lenton & Latour, 2018) in which integrated organisms, especially bacteria, protists, fungi, and plants shape the whole biosphere through direct and indirect influences on the atmosphere, pedosphere, and even the geosphere. Indeed, the fungal hyphae and plant/tree root network act in a similar fashion to the internet which was only invented by our technological civilization a few years ago. The important difference is that the natural 'wood-wide web' (Sen, 2001; Giovannetti et al., 2006) was invented by living cells some 500 million years ago, representing a living and evolving system composed of sentient cells.

Anaesthetics and Plants: From Cytoplasmic Streaming Up to Traps of Carnivorous Plants

Plants have served as valid objects of study for investigations of anaesthetics ever since their anaesthetic pain-relieving properties in humans were discovered in 1846 (as was discussed in more detail in Chapter 10). Here we focus on two aspects, their effects on cytoplasmic streaming in plant cells and their impacts on carnivorous plants hunting insects. Some large plant cells provide an excellent experimental system for studying cytoplasmic streaming and contractility. Eduard Strasburger used this topic as his signatory one in his rectorial speech at the University of Bonn in 1891 (Strasburger, 1891; Volkmann, 2013). Alfred Ewart published an extensive review on this topic in 1903 (Ewart, 1903). He credited Willy Kühne as the first to report that ether and chloroform arrest cytoplasmic streaming in the large stamen cells of *Tradescantia virginica* in a manner similar to that when they are exposed to electric fields (Kühne, 1864; Ewart, 1903). Interestingly, Kühne was a student of Emil du Boys-Reimond who discovered action potentials and initiated the field of experimental electrophysiology. Others confirmed that anaesthetics not only block the movements of supracellular organs, as discussed below, but also arrest subcellular movements in cytoplasm (Osterhout, 1952; García-Sierra & Frixione, 1993).

Since their discovery, anaesthetics have intrigued researchers for their ability to not only block the sensory awareness of organisms, including pain, but have similar effects on the movements of their organs, causing general immobility (Seyfarth, 2006; De Palma & Pareti, 2011). Importantly, action potentials are linked to cytoplasmic streaming as there is a sudden cessation whenever an action potential is fired (Ewart, 1903; Baluška & Mancuso, 2019). It has been proposed that the subcellular actions of anaesthetics on excitable membranes are critical in these supracellular effects on plants and animals, including humans. In all organisms, anaesthetics cause immobility via impacts on the excitable plasma membrane, blocking action potentials and organ movements. In plants, the best studied organs are the leaves of *Mimosa pudica* and leaf traps of carnivorous plants (Yokawa et al., 2018; Scherzer et al., 2022). Obviously, long-distance bioelectric signalling has the same biological features in plants and animals. It is essential for the movement of organs, proceeding via the same membranous anaesthetic targets as those that block subcellular cytoplasmic streaming.

The N-Space Episenome of Plant Bodies: A Central Commander and Centre of Plant Self-Identity

As we have discussed in Chapter 6, bioelectric activities of the excitable plasma membrane generate bioelectric fields around themselves which extend across the inside of the cell and radiate outward. In his recent book *The Song of The Cell*, Siddhartha Mukherjee (2022) reminds us of the five basic tenets of life. To complement them, the CBC theory adds a sixth and seventh, thus providing us with the Cell Theory of Life:

1. All living organisms are composed of one or more cells.
2. The cell is the basic unit of structure and organization in all organisms.
3. All cells come from other cells.
4. Normal physiology is a function of cellular physiology.
5. Macro-organic disease, as a breakdown of physiology, is the result of disrupted cellular physiology.

6. In addition to the genome and epigenome, every cell is endowed with a senome that underlies cell sentience.
7. Sentience and consciousness of all multicellular organisms derives from senome-based cellular sentience.

Tenets 1 and 2 were proposed by Theodor Schwann and Matthias Schleiden; tenets 3, 4, and 5 come from Rudolf Virchow. We are adding tenets 6 and 7 to complement these and, in so doing, complete the definitions of imperatives of cellular life. As all organisms are based on one or more cells endowed with cellular sentience, it is only logical that all multicellular organisms are endowed with an existential consciousness that permits their navigation through life and survival. These seven tenets, then, are the essential precondition for biological evolution as it has been sustained for more than 4 billion years. Notably, tenets 6 and 7 suggest that the senome of the cell is the basic unit of mental life. Plant cells are more complex than animal and human cells due to their chloroplasts, representing a second endosymbiotic organelle. This status means that the metabolic and senomic integration of plant cells is more complex than animal cells insofar as they ultimately require more integrated components.

As discussed in Chapter 9, we have proposed that senomes (Baluška & Miller, 2018) merge together and assemble N-space Episenomic fields (Miller et al., 2020a, 2020b) which guide organismal life and evolution. The evolution of such integration was very slow which explains the long timeline in the evolution of the eukaryotic cell as a combination of several pre-existing prokaryotic cellular life forms, thereby permitting their further evolution as true multicellular organisms within the new domain, Eukaryota. Plants do not possess a centralized neuronal system like the central nervous systems in animals. However, their requisite cell-based N-space Episenome could explain how they can generate self-agency, sociality, and cognition even without a central neuronal organ (Baluška & Mancuso, 2009a, 2009b, 2020, 2021).

From this perspective, the N-space Episenome of plant root–fungal hyphae networks of the wood-wide-web is the largest and oldest episenomic field on the planet Earth. It started on its evolutionary path with the first water-based plants and fungi moving to land some 400 million years ago. As it is of central importance for the Earth's climate (Baluška & Mancuso, 2020), these same roots are also essential for the future of our civilization.

12

Issues of Ethics and Morality

Entailments of the CBC

Introductory Remarks

One of the central themes of this book is that our CBC model and its associated research strategy differs in fundamental ways from what we have been calling the 'standard approach'. We begin the exploration of sentience and consciousness with the simplest life forms and follow the evolution of cognition as it developed in more complex species. As noted in several places, the more common framework is the one that begins with the mental life of *Homo sapiens* and looks backward through the evolutionary tree for behavioural commonalities, biomechanical analogues, and homologues of sentience. Working within the standard framework, the approach to issues of ethics, animal welfare, imposition of suffering, the guidelines and regulations of medical research, commerce, and food processing is relatively straightforward. The efforts turn on identifying the species that are determined to be sentient, feel pain, and experience suffering and establish guidelines for their welfare. As we will see, problems have emerged within this framework primarily because different researchers and advocates for welfare identify different species and clades as ones that do, in fact, suffer, do feel pain, and should be given consideration in all aspects of relevant commercial and research enterprises.

The CBC, on the very other hand, invites, or better, insists that a variety of additional issues be put under the microscope. We will try to cover all that seem relevant, noting that in most cases we have no definitive conclusions, only a motivated framework within which to operate. Our hope here is to engage in and encourage discussion, debate, and bring in an alternative point of view. We have a conceptual platform, it tends to be inclusive, but pragmatic and emphasizes the dignity of all life forms, the interconnectedness of individual organisms and the planet, builds on efforts to preserve the ecology of complex biospheres, and outlines a general framework for how to function ethically, behave morally, do as little damage, cause as little pain as possible, and work to mitigate that which results from actions taken and decisions made.

The Sentient Cell. Arthur S. Reber, František Baluška and William B. Miller Jr., Oxford University Press.
 DOI: 10.1093/oso/9780198873211.003.0012

These are issues that have not, so far as we have been able to discern, been broached in the literature but, in our view, need to be. As we have noted in several places (Baluška et al., in press; Reber et al., in press), the *Homo sapiens*-focused strategy is non-optimal and is the reverse of that taken in the physical and chemical sciences. There, the standard approach has been to begin with the simple and work towards more complex circumstances as insight and understanding develop. A critical feature of our CBC model is that it emulates the natural sciences—begin with the simplest species. In this final chapter, we will examine the manner in which these two research strategies deal with matters of morality, ethics, and species welfare.

Let us begin this exploration with a simple fact about the carbon-based life on this planet—one that some might find uncomfortable:

> Except for unicellular prokaryotes and photosynthetic organisms, who are content to consume sugars and minerals, every organism on the planet lives by ingesting other organisms. Ingesting, of course, is made possible by engaging in acts that damage or kill other organisms—with a few species that wait for others to do the deed and then come in for feeding.

In the following discussion we will raise issues that have been part of a variety of ongoing conversations in fields as diverse as moral philosophy, evolutionary biology, medical research, food science and commerce, species welfare, and the role of governmental regulation. In addition, there are issues that have significant personal features in lifestyles and still others that make contact with larger social and cultural deliberations. Our goal here is to focus on the implications of extending these issues to other biological domains examined in a framework based on evolutionary biology and the deep message of the CBC model.

The Standard Model Versus the CBC Model

Those who work within the standard model have had ongoing discussions on a variety of ethical issues but the efforts have led to a crazy-quilt array of varying opinions on pain, animal suffering, and animal welfare. One can see this pattern in a recent overview of the issue by UK psychologist Matan Mazor and colleagues (Mazor et al., 2022) who examined the link between consciousness and ethics. They noted that for centuries the central point has turned on whether one believes that the species under consideration has a mind, and perforce, must be thought of as conscious. Perspectives have run the gamut from Aristotle who argued that only fully rational humans had an existential consciousness and that concerns about other species were not an issue, through Kant's position that only 'autonomous' organisms were so mentally endowed, and Bentham's focus on whether they could suffer and experience pain. Mazor et al. reviewed the many conclusions drawn by philosophers, psychologists, anatomists, entomologists, primatologists, and neuroscientists. Not surprisingly, all have taken the 'standard' approach and the search is for behaviours and/or neural structures that are sufficiently similar to those observed in humans that the species can be considered to have a palpable consciousness and thereby have valenced experiences and, critically, can suffer.

We maintain that all these varied approaches and the mutually contradictory conclusions they have come to are simply missing the point: life and sentience are coterminous so, yes, all species including the simplest unicellular prokaryotes, suffer when experiences with negative valence are present. As noted in the review of prokaryote behaviours in Chapter 2, unicellular species learn to avoid places where aversive chemicals are present, locomote away from temperatures that are dangerously low or high, block entry of molecules that they detect might cause cellular damage, and communicate distress at lowered nutrient levels—all of which support the conclusion that they are sentient, have valenced experiences, and are capable of suffering. At the cellular level, suffering need not equate with our specific human context. As has been explained, cells have a well-defined state of preferential balance. When perturbed, cells rapidly mobilize their limited resources to correct that loss of essential equipoise to restore their preferential state. We humans do the same at our scale. Further, as we discussed in Chapter 11 on the sentient life of plants, flora also display a host of behaviours and biomolecular functions that are compelling evidence that they too are conscious beings and aware of aversive states. So, unlike those who have clung to the 'standard model', we have to confront the ethical issues that these data raise from a very different footing.

Utilitarianism

The philosophical model that lies behind many of the discussions about welfare of species is *utilitarianism* or its earlier instantiation, *consequentialism*. There have been a variety of different forms of utilitarianism put forward over many centuries, but in all, the guiding standard is to try to ascertain the ultimate consequences of actions, to determine to what extent decisions and operations based on them increase overall *utility*—where utility is treated as a general goal and applied across all the 'individuals' impacted by the decisions and actions. In short, choose actions that have the least negative impact on the species under consideration and, of course, those that maximize satisfaction and happiness.

There are a number of variations on utilitarianism but each holds onto this foundation principle. Actions and decisions should be judged on the basis of which choices maximize happiness, satisfaction, and wellness for the most 'individuals'. In the standard interpretations, the 'individuals' in the analysis are people.[1] In more recent approaches, the coverage extends to 'sentient beings' and, again we find the usual struggles to identify which species to include and which choices and decisions should be made to provide for maximum welfare for them. We are comfortable with this broad and non-technical utilitarianism—with the understanding that it does not resolve many of the problems that emerge when decisions need to be made. Its advantage is that it provides a pragmatic framework within which to operate.

British moral philosopher Phillipa Foot, a critic of utilitarianism, introduced a thought experiment some years back known as the *trolley problem* in which a runaway trolley is barrelling down the tracks and heading towards five people who, in the original version, have been tied to the tracks by a mad philosopher. However, there is a switch you can pull which will divert the train to a siding. On the siding, however, is a single person, also tied down. Do you pull the switch? Do you act in a manner that brings about the death of an individual but saves the lives of five others? The dilemma has been so thoroughly studied and so many variations on the theme developed, that there is a subfield in moral philosophy known as 'trolleyology'. It is easy to see how the situation can be made more complex and other factors play a role. For example, in one variation the single individual is a large, obese person on a bridge over the trolley tracks. The only way to save the five persons is to push the obese person over the edge of the bridge—an act very different from simply pulling a switch. In another, the five are all convicted child rapists and the single individual a respected jurist. Now, how do you deal with the situation? Put your father or your child on the single-person track. Et cetera.

In the more familiar terrain of the CBC, we can have a variation on the problem where the entities are sentient unicellular organisms and the setting is a Petri dish where a stream of antibiotics is heading towards five prokaryotes but can be diverted to the other side where there is a single cell. If the five are all *Mycobacterium tuberculosis*, which are deadly, and the single cell is *Lactobacillus acidophilus*, which promotes health, the choice is easy. Flip the

[1] Wikipedia has a useful entry that touches on the various utilitarian approaches that have been put forward and the various critiques on them (https://en.wikipedia.org/wiki/Utilitarianism).

species and the decision changes. Note that the taking of life of equally sentient organisms ceases to be the critical feature. While the emotional impact of deaths of the organisms in the Petri dish version is far less than when human beings are involved, there is little doubt that we would prefer to kill as many deadly bacteria as we can and protect the lives of beneficial ones.

But in the final analysis, the deep problem with utilitarianism turns on the difficulty, if not impossibility, of ascertaining what the ultimate consequences of actions will be. The trolley problem is a nice example of what philosopher Dan Dennett calls an 'intuition pump' in that it stimulates critical thinking in an area or on an issue but it does not lead to any satisfying conclusions. There are, in most complex situations where the principle is raised, just too many unknowns and efforts to make decisions about actions and choices based on what are often simply educated guesses about the future impact of decisions. In short, we have adopted a *pragmatic* utilitarianism. We appreciate the importance of analyses of the ultimate outcomes of choices and actions but appreciate the criticism that in most circumstances the final decisions will be lacking in ultimate utility.

The Precautionary Principle

The precautionary principle is a broad ethical standard that is tucked into utilitarian thinking and calls for decision-makers to take into account the possible harm that might ensue if a particular course of action is chosen. It gets invoked in a host of circumstances, such as considerations about whether novel medicines or treatments should be provided to the public, decisions about which species can be used in medical and biological research, and whether new inventions or engineering techniques that might have an adverse impact on the environment should be adopted. In the context of the topics we are considering here, the relevant concern is whether it should be a factor when decisions are made about the welfare of various species and how we should treat them.

British philosopher Jonathan Birch, whose approach to animal sentience was discussed in Chapter 1, argues for establishing a conceptual bar for when to apply the principle and it is, not surprisingly, based on the determination of sentience and the experience of pain. When, Birch maintains, there is compelling evidence that a species or family is sentient and feels pain, then caution should be taken when dealing with its members and regulations should be put in place that protect them from unnecessary harm or suffering (Birch, 2017). Species that are determined to be insentient, or ones where there is insufficient evidence that they are, do not need to have any precautionary guidelines in place. Australian economist Yew-Kwang Ng (2017), a supporter of the precautionary principle, hedges when the discussions turn on species he regards as non-sentient. When, he maintains, there is no reason to conclude an organism is sentient then the issue of welfare is not relevant because those species simply have 'no welfare'.

Several of those who wrote commentaries on Birch's paper pointed out that there are other considerations that should play a role in the decisions. As in the discussion above on the trolley problem, should we distinguish between species that are detrimental to our lives and those that are beneficial? Mosquitoes are the only species that kill more humans than humans. Do they deserve to be granted precautionary treatment? How do we feel about the five species of bacteria that collectively are the second largest killer of humans causing fully one in eight deaths (Ikuta et al., 2022)?

It is not hard to see the problem that the CBC stance has to deal with. One of us (Reber, 2017) wrote a comment on Birch's paper arguing that, since all species including unicellular prokaryotes are sentient, all species should be subjected to the kinds of concerns that the precautionary principle insists on. But there are complicating factors in play. We will quote a section from Reber's commentary as it lays out the issues that are opened by our theory:

> Then there are those sentient bacteria. Some probiotic species, like *Lactobacillus acidophilus*, promote health; others like *Mycobacterium tuberculosis* are deadly; and still others like *Escherichia coli* have strains, some beneficial, others decidedly not. Should only some come under animal protection legislation when presumably all share equivalent mental states? We routinely tolerate the imposition of pain when a goal deemed worthwhile is involved.

As noted above, the debate has been going on literally for centuries with most—scientific researchers, philosophers, and government regulatory bodies—concluding that there should be some forms of oversight to prevent or mitigate suffering when a species is considered to have an existential consciousness. A classic example of the kinds of issues raised can be seen in the ongoing discussion involving fish pain. The opening salvo was a paper published in the online journal *Animal Sentience* by Brian Key (2016) titled 'Why fish do not feel pain'. The paper has received over 40 (as of late 2022) published commentaries and Key has responded in three separate essays.

Key's primary claim is that fish do not feel pain because they lack the cortical structures that are known to be implicated in human experiences of pain. Some, such as biologist Georg Striedter (2016), argued that the lack of a neocortex does not imply a lack of painful experiences and maintains that there is not sufficient evidence to come to any conclusive answer. Others, such as Stuart Derbyshire (2016), sided with Key. One of us (Baluška, 2016) commented that there is compelling evidence that plants react to events that cause cell damage by producing chemicals that have painkilling properties and, importantly, as discussed in Chapter 10, are sensitive to anaesthetics. If plants feel pain without anything resembling a mammalian cortex, it is almost certainly the case that such neural structures are not essential and that fish do, in fact, feel pain.

Stuart Derbyshire (2016) began his commentary with a sentence that makes clear where the problem is and why there is even a discussion about whether or not fish feel pain: 'Debate about the possibility of fish pain focuses largely on the fish's lack of the cortex considered necessary for generating pain.' Again, we see how the 'standard model' of starting the explorations with human experiences and human neuroanatomy throws the issue into totally unnecessary confusion. The mental states of *Homo sapiens* who are suffering are embedded with human mental representations precisely because of the ways in which our nervous systems are wired and how the various cortical and subcortical centres and pathways deal with the environmental inputs. But the conclusion that fish *do not* feel pain is absurd. Do they feel pain differently than we do? Does the lack of a neocortex have an impact on the kinds of pain they experience? Doubtlessly. They also respirate differently, locomote differently, process the visual spectrum differently, and detect odours differently. Piscean species and mammals evolved on distinct branches of the evolutionary tree. But it is misleading in the extreme to conclude that, because their brains are wired differently from ours, that they lack certain cortical centres and pathways, that they do not actually feel pain, and do not suffer when their bio-physiological beings are under stress. The fish flopping about helplessly in the bottom of a boat with a fishing hook piercing its mouth is drowning and is doubtlessly in considerable pain.

Similar arguments have been put forward by bio-ethics philosophers Mikhalevich and Powell (2020) who point to the cognitive functions of invertebrates and argue that they should have the same kinds of standards of treatment as vertebrates. As they put it, excluding invertebrates is 'motivated by cognitive-affective biases that covertly influence moral judgment.' In a similar vein Robert Jones (2013) presented an overview of the various ethical considerations that arise when animals are used in industrial food production and preparation as well as in scientific research. Not surprisingly, both efforts focus, as we have seen consistently, on multicellular, larger species and use evidence of sentience as the criterion for recommending concerns about well-being and the imposition of government regulations. Once again, we appreciate their efforts but view them as limited. The issues are larger and pertain across the evolutionary tree. As we outlined in detail in Chapter 10, there is virtually no doubt that all living organisms suffer and feel pain as negatively valenced experiences in some form and note that all respond to anaesthetics in biomolecularly similar fashions—all of which function to mitigate the unpleasant experiences. What is the point, evolutionarily speaking, of a species being sensitive to and responding in appropriate ways to anaesthetics if it did not experience something akin to what we call 'pain?' It would be a waste of biomolecular resources for no gain—which is not an adaptive evolutionary strategy.

Welfare—The Need to Diminish Suffering and the Planetary Partnership

Recognizing that all living organisms can feel pain, can suffer, makes it clear that we oppose the stance that Key and Derbyshire cavalierly take and support making species welfare a central concern. Our position is that all species should be treated with dignity with the understanding that the rights and needs of other species are held in the balance. From our framework, since all living organisms share fundamental biomolecular functions, we are all in a planetary partnership, one that calls for a reciprocal relationship between parties. Maintaining a broad, interlocking affiliation with others involves acknowledging the dignity (self-identity) of each other. Planetary biology is the narrative of the maintenance of conscious self-reference. The CBC model emphasizes planetary stewardship and obliges fully embracing all other creatures including plants as partners in the fullest sense of the word. When we dispose of living planetary resources, our duty should be to try our best to maintain or strengthen the planetary connectedness of a living partnership.

But, there is no question that our CBC stance opens a can of worms and all of its ramifications have to be addressed. Pertinently, there are only two things we securely know: (a) all cells are conscious, hence all multicellular organisms are as well; and (b) all of evolution is a consequence of that reality as co-evolution and co-development across all species. However, admittedly, many of the particulars of intelligent life remain opaque. Pain among different organisms is a good example, which is one reason why it has become the focus of much debate. The fact remains that we do not know which organisms feel pain in the sense we understand it since pain, just like the wavelengths of perceived light, is differently sensed among organisms. But, of course, if a plant does not feel pain, our kind of pain, it is certainly stressed by events that threaten its existence or well-being—as was discussed in Chapter 11. This experience contradicts its state of preference, can be categorized as unpleasant or negatively valenced, and, because of this subjective state, requires the mobilization of resources to mitigate it. However, we need to keep in mind that things in the world of flora are not simple. Many plants thrive by adroit pruning, by the removal of branches and limbs. For example, the hardy oleander, a plant commonly used as a living fence, flowers much more lushly when pruned annually and, interestingly, actions that limit its growth produce a healthier plant that is more disease and pest resistant. Unpruned oleander looks distressed with many dead vines and areas of diseased segments. Other species, such as the prickly pear cactus, also grow more vigorously through self-pruning. Such traits suggest that these mechanisms and others we described in the previous chapter emerged over time simply because they had evolutionary advantages.

We are, as yet, unprepared to make too many judgements of 'better' good for many types of living organisms. Cutting grasses might be a negative for the grasses but sends more sunlight to the level of the earth, which brings up worms, which attracts birds, that fertilize the soil. No decision about what to do with plants is only about plants. It is encased in an entire local ecological frame, with complex further ramifications. This issue, of course, is not new. *Mad Magazine*, one of the longest running publications of cartoon humour (1971–2018), once featured a parody of the Flash Gordon character who was the hero in a limited series of comic books published by DC Comics and the protagonist in a widely trashed film of the same name. The characters in the story had landed on a planet which, to their astonishment, had mud people. If they stepped on mud, they realized that they were hurting the mud people. The joke was that they realized they were trapped because it turned out that there were also air people, rock people, and so on. Everywhere they went, something said 'ouch'. In the end, they decided that if you are going to hurt something no matter how or where you move, just get moving.

We do not know the answers and cannot offer any specific conclusions. Again, our role in this chapter is to initiate debate—with the understanding that it takes place within the CBC framework. All organisms are conscious but we have been gifted special planetary privileges due to our unique human conscious frame of reference that links to unparalleled engineering and problem-solving (and creating) capacities—although we eschew the pan-psychism that the Flash Gordon parody reflects. In our theory, only biologically based organisms are sentient and can feel pain. Such privileges carry countervailing responsibilities. Our planetary dominance obliges thoughtful planetary stewardship.

Stress and Negative Experience in Plants

The CBC framework opens up other related concerns and, as we hinted above, understanding the role of flora is one. Because all cells are conscious, all multicellular organisms are as well. And, as noted in several places, evolutionary mechanisms rarely operate by jettisoning functional and adaptive traits and sentience is about as adaptive a characteristic as there is. In short, it would be extremely unlikely if the millions of species in the plant world were to have evolved without the capacity to experience stress as negative and not to have biomolecular functions to mitigate it. Having said that, we recognize that many of the particulars of intelligent life in plants remain opaque but need to bring up a few of those 'intuition pumps'.

1. Let us return to the one we touched on earlier. Does grass hurt when you mow it? Interestingly, this is an oft-asked question on social media where the answer is almost invariably 'no' but without any reason why. Like a lot of other plant species, grass, being a member of a clade of monocots, likely evolved to cope well with being eaten, or in modern times, mowed. The situation is similar to that in fruits, nuts, berries, and a host of other plant species that serve as another organism's food. In all cases, it is an effective way for sessile species to distribute their genes to other locales which, depending on the symbiotic relationship between the species, can be very far away. Many bird species, especially migratory ones, fly immense distances. Similarly, many species of ungulates travel in herds over great areas in their search for sustenance. Hence, these granular species likely evolved so that there is no suffering when some of their organs, tissues, and cells are cut or eaten.
2. Here is another touched on briefly above. Are plants stressed when pruned? As noted, it is likely that there are many that are not. In many species of bushes, shrubs, and trees, pruning promotes healthy growth. It is also best done at the collar of the offshoot branch. Trees and other plants that invite pruning evolved antibiotic substances that are concentrated at the collar and operate to prevent or reduce the likelihood of an infection. The implication is that a mechanism evolved that functions to increase the chances of a plant's survival, even if it appears on the surface to be one that would, in others species, produce tissue damage

and suffering. It is also apparently the case that the point where the off-branch emerges is structurally weaker, making this the point most likely to be break off under naturally occurring conditions. Again, this would have evolutionary advantages and would have emerged well before pruning scissors were invented.

3. Generally speaking, is there a need to address the larger question of welfare in flora? Since the CBC theory argues that the functions that emerged in prokaryotes were/are carried across all subsequent species, the capacity to experience stress as negatively valenced, to suffer, goes along for the full evolutionary ride. The reason this gets complex is because the current approach to animal welfare, as noted above, turns on whether the species under consideration are sentient.

Recently, writer Matthew Rozsa ventured into the field in a non-technical essay in the popular magazine *Salon* (Rozsa, 2022). While doing due diligence as a reporter and writer, he focused on our work and the CBC model noting that 'plants seem to possess a different set of evolutionary tools that suggest they may experience consciousness, albeit in a radically different way from us'. He also interviewed one of us (Baluška) and a colleague, Spanish philosopher of science Paco Calvo, and reported that:

> 'We should acknowledge that plants are complex living systems which deserve dignity, as it is stated in the Swiss Constitution through amendment from 2008 (see Harmon, 2009)', Baluška argued. 'As animals, humans and plants are in close co-evolution and have the same biological origins, we should treat them as living organisms deserving dignity.'

Calvo noted that, even if humans only acknowledge that plants have a very primitive form of consciousness, that should still make us feel 'uneasy' at the realization that 'plants are agents, and not mere objects or resources to be exploited more or less wisely'.

In short, plants should be dealt with in the same manner that most ethicists have argued other sentient species should be. Use the precautionary principle to prevent unnecessary stress and recognize that they are part of the fully balanced environment within which we all live. Apply a pragmatic utilitarian position in an effort to recognize what the implications of our actions are and try to mitigate what suffering we do cause. We began this chapter with the observation that all species, other than prokaryotes and photosynthetic eukaryotes, can survive only by damaging or killing and consuming other organisms. It is simply not possible to survive without causing stress and pain but it can be handled with grace and sensitivity and an acknowledgement that we are all encased in a complex, interlocking biosphere in which all are worthy of being treated with dignity—and, yes, in a complex ecologically structured world, this includes those who are, shall we say, not our best friends when it comes to our own well-being. Of course, in households with the pragmatic utilitarianism values we are promoting, insects like ants, bees, and spiders who venture indoors are gracefully shown the door and told to handle life as best they can. Mosquitoes and cockroaches are swatted with no regrets.

Vegetarianism and Veganism

So far we have worked around a topic that sits up front when the issue of plant sentience is raised: what are the implications of the CBC model for vegetarians and vegans? There are many reasons for adopting a vegetarian or vegan diet; these include promoting health, the previously raised issues of animal welfare, concerns about the treatment of sentient species in industrialized food processing and science, worries about the excessive use of environmental resources, religious convictions,[2] dislike of the 'mouth feel' of meats, and, of course, ethical considerations that arise from the stress and pain that the animals that provide sustenance undergo. For vegans, all these considerations play out as well with the additional feature that no animal-derived foods are eaten so that eggs, dairy products, and gelatin are not consumed.

But these concerns are, once again, applied to what are thought to be sentient species and as we have seen repeatedly, plants are not regarded as sentient and, as a result, generally not accorded considerations of species welfare. The CBC, as we have noted countless times, changes the way this issue must be viewed. All cellular life is conscious, senses external events, experiences and evaluates internal conditions, thinks, makes decisions, and has negatively valenced sensations and perceptions. Is there a way for vegetarians and vegans to adapt to this compelling reality—particularly those who have adopted their lifestyle for ethical reasons? There is, and it is obvious. Restrict your diet to those plants that want to be eaten, those whose evolutionary history is marked by the development of biomolecular mechanisms that render them tasty and nutritious to other species and did so for fundamental evolutionary reasons. Being edible was an excellent strategy for sessile species to evolve as it supports

[2] We note that many of the points we are making here are very similar to the foundation principles of Jainism where the path to enlightenment calls for non-violence and harm reduction to all life forms, including plants.

the distribution of their DNA to areas other than their immediate location. These include virtually all the floral forms that we have touched on: grains, rices, berries, fruits, nuts, legumes, beans, root vegetables, leafy greens, herbs, and many spices.

It is also worth keeping in mind that plants and other species co-evolved and have, over time, developed modes of communication networks with plants that we and other species eat that are part of our epigenetic adjustments to environmental cues. For example, exosome-like nanoparticles that are ingested from ginger, carrots, or grapes can be absorbed by macrophages and affect mammalian gene expression. We share a tightly linked co-evolutionary connection with all participants in our relevant ecosystem, since we are contributing back our cellular DNA and microbial partners to plants.

Altruism and Its Place in Evolution

As discussed in Chapter 2, prokaryotes display a form of altruism. When cells in the middle of a colony become nutritionally stressed, they communicate their metabolic condition to others in the biomass who adjust nutrient uptake and cell division until the nutritional deficits are mitigated. We have called these demonstrations 'altruism' because they have the criterial features that have been used when referring to similar behaviours in other species. The cells on the periphery, the ones that reduce cell division and nutrient uptake, are putting themselves at risk to benefit their colony-mates. Being on the periphery of a biofilm means being in contact with predators, viruses, shifts in the chemistry, and physical properties of the environment. Reducing cell division and nutrients is risky. Being in the centre is protection from these dangers but, of course, there is a limited food supply. See Humphries et al. (2017) and Prindle et al. (2015) for details on these mechanisms and how they operate and note that Humphries and colleagues observed these effects when they took place between different species and different colonies—an example of cross-kin altruism.

These effects, which molecular biologist Gürol Süel dubbed 'time sharing', are examples of the kinds of ecological integrity that is characteristic of all life. When displayed by relatively simple unicellular species, the full expression of the interactive N-space Episenome is, of course, going to be relatively simple but, in keeping with our overall theme, these fundamental biomechanisms emerged with the first life forms and became the organic platforms for all of evolution. We are not saying that these *ur*-forms of altruism that unicellular prokaryotes display are accompanied by self-expressed feelings of having done something noble or a sense of thankfulness for the efforts of their conspecifics. But we are saying that they are experienced as valenced states of awareness.

Sustaining selfhood of each ecological environment should be considered an explicitly positive ethical goal. The larger question of ethical conduct remains an individual one but, once the sentience of all organisms is recognized, we are all jointly tasked to limit our 'taking' from any ecology and to take care to remain within the limits of sustaining ecological partnership relations. Putting oneself at risk for the potential benefit of others is deeply embedded in our evolutionary history. It is part of the partnership, the ecological, social links that bind all members of whatever community they live in, be it a sophisticated functioning society or a Petri dish. Every organism is a discrete self with its own character. Every ecological setting also has its distinct pattern of combinatorial self-hood. This way of framing the issues reflects the principles discussed earlier in Chapters 3 and 9 where we identified the importance of a cellular pervasive information field and N-space Episenome. Self-hood is, at its core, a reiterating phenomenon at differing scales. Holobionts and unicellular organisms combine in ecological systems that support the lives of all participants.

It is worth noting that phrasing the larger issue of moral functioning in this manner is a form of consequentialism in that it points to a philosophical stance that emphasizes actions that, in ideal settings, bring about the optimal end-states for those settings. See moral philosopher Douglas Portmore's book *Commonsense Consequentialism* (Portmore, 2011) for a detailed analysis of this perspective. Recently *The Guardian* had an article reflecting on the Law Society's argument that legal rights should be extended to animals, plants, and even rivers on grounds similar to what we and Portmore are proposing (Siddique, 2022). Such regulatory shifts would promote biodiversity, mitigate the impacts of climate change, and have generally beneficial effects on the planet. Several countries, including Ecuador and Bolivia, where climate change has had a significant impact on the ecology, have embraced the principle by making 'ecocide' a prosecutable offense.

The Final Word

Well, there really is no 'final word'—not when adherents of the CBC model such as us confront the full extent of its entailments. We have raised these issues here primarily to promote discussion and to tune thinking towards the notion of a participatory dignity that we believe all should embrace. Make decisions within the scope of a pragmatic utilitarianism, use the precautionary principle when doubt is encountered, do as little damage to others (all others) as you can, and cause as little stress as possible as you live your lives. It is what we try to do.

APPENDIX I

An Exercise in Lexicography: Defining(?) Consciousness

As noted in the Preface and sprinkled throughout the book, there were reasons for taking the non-definitional route when dealing with our core term, *consciousness*.[1] One of them was what we found while doing our due diligence, searching for a useful definition of our central concept. A reasonable place to start the exercise, we figured, was to look for 'official' definitions of the term in the peer-reviewed reference literature. The most 'official' place we know that fits this description is the Penguin Books *Dictionary of Psychology* and, since one of us, Arthur Reber, wrote it, we felt comfortable in starting there. In the first edition, which was published in the middle 1980s, back when cognitive psychology was emerging from the hinterlands and rapidly becoming the dominant approach in psychology, *consciousness* was given there as:

> 1. Generally, a state of awareness; a state of being **conscious**. This is the most general usage of the term and is intended in phrases like 'he lost consciousness'. **2.** A domain of mind that contains the sensations, perceptions and memories of which one is momentarily aware; that is, those aspects of present mental life that one is attending to (see **attention**). **3.** That component of mind available for **introspection**. This meaning is found in the older writings of structuralists and other introspectionists. **4.** In psychoanalysis, the **conscious**.
>
> The term has a distinctly checkered history. It has been represented at times as the central focus of psychology (e.g., **structuralism**) and at others has been banned from the psychologist's lexicon as representing nothing more than the epiphenomenal flotsam of bodily activity (e.g., **behaviorism**). The ongoing fascination with it, however, stems from the compelling sense that consciousness is one of the defining features of our species: that to be human is to possess not only self-awareness but the even more remarkable capacity to scan and review mentally which we are aware of. [Note: the bolding here is not for emphasis. It simply signals that the term is defined elsewhere in the dictionary.]

The entry went on to discuss other connotative issues but this gives us a pretty good feel for how the term had been used in psychology. It did the job and touched most of the issues, but it didn't do justice to the range of issues currently under discussion. The entry was also entirely species-centric. It dealt only with human consciousness and treated the term as though it denoted functions and processes only found in *Homo sapiens*. What was being reflected, of course, was the bias of the field. Twenty years, three editions, and two new co-authors later, the fourth edition (now with Emily Reber and Rhianon Allen) still displays these same themes. There is more up-to-date material but the definition still reflects a species-centric stance. If there ever is a fifth edition, this narrowness of vision will be changed.[2]

How about our chief competitor, the *Macmillan Dictionary of Psychology*? The entry on consciousness was written by lexicographer Stuart Sutherland. Here, there is equally little assistance. Sutherland wrote:

[1] Some of the material in this appendix appeared in Reber (2017). It is reproduced here with permission from both Reber and the publisher, Oxford University Press.

[2] Penguin has let us know it is extremely unlikely that there will be any future editions. The reason? The obvious, the Internet.

> The having of perceptions, thoughts, and feelings; awareness. The term is impossible to define except in terms that are unintelligible without a grasp of what consciousness means. Many fall into the trap of confusing consciousness with self-consciousness—to be conscious it is only necessary to be aware of the external world. Consciousness is a fascinating but elusive phenomenon; it is impossible to specify what it is, what it does, or why it evolved. Nothing worth reading has been written about it.

Sutherland's approach rather made it seem that consciousness was like a cold. You cannot really understand what it is like or how you feel unless you have had one. Clever but not terribly illuminating—although Sutherland's tilt toward the subjective, experiential aspect was typical of the way most used the term. That last line was, for those who knew him, classic Sutherland. It also made his easily the most frequently quoted definition.

These two efforts came from psychologists with a cognitive orientation. It is possible that psychologists' efforts might prove wanting as we tend to have our own views on things mental. So let us turn to the authoritative, eight-volume *The Encyclopedia of Philosophy* compiled and edited by another long-gone old friend of one of us (Reber), philosopher Paul Edwards. Paul reached out to philosopher Charles Landesman to handle the entry on consciousness and Landesman promptly provided more heat than light:

> The term 'consciousness' occurs in philosophy, psychology, and common speech with a variety of different meanings. In this article those meanings will be discussed which are relevant to the formulation of several recurrent philosophical problems.

Unfortunately those 'meanings' turned out to be categories of reference which allowed Landesman to duck all definitional demands. He started out treating consciousness as though it could (should?) be thought of as self-knowledge and then provided a historically sensitive overview of this issue with attendant discussions about introspection, mental states, and whether or not these states of consciousness are made up of separate, non-material 'stuff' (i.e. the 'Hard Problem' back before it was so named). It wasn't that Landesman's entry wasn't scholarly. It was. But stripped bare of the many segues, what was left was not much more than the folk psychology connotations common in the field and it certainly did not give us a definition.

The next place checked was the entry in Simon Blackburn's respected *Oxford Dictionary of Philosophy*. Alas, it was a lot like Landesman's. When thrown a hard slider up and in, Blackburn also ducked, bailed right out of the batter's box. His definition wasn't a definition at all. The entry started out:

> Possibly the most challenging and pervasive source of problems in the whole of philosophy.

And then he described and outlined those problems without making any attempt to define the term itself.

But lexicographic cowardliness of this kind is nothing compared with the *Penguin Dictionary of Philosophy* edited by Thomas Mautner. You might think you or someone had made some kind of alphabetic sorting error but search as you may you will not find an entry for *consciousness*! And nothing for *unconscious*, or *experience*, or *awareness*. Fascinating how a term regarded as 'the most challenging and pervasive source of problems in the whole of philosophy' by Blackburn can be utterly neglected by Mautner.

Beginning in 1995, Stanford University began publishing, editing, and regularly updating the freely accessible *Stanford Encyclopedia of Philosophy*. It has become one of the most respected online sources for general information on concepts and terms in the field. The entry on consciousness begins in a manner that by now those of you still with us here should be ready for:

> Perhaps no aspect of mind is more familiar or more puzzling than consciousness and our conscious experience of self and world. The problem of consciousness is arguably the central issue in current theorizing about the mind. Despite the lack of any agreed upon theory

> of consciousness, there is a widespread, if less than universal, consensus that an adequate account of mind requires a clear understanding of it and its place in nature. We need to understand both what consciousness is and how it relates to other, nonconscious, aspects of reality.

What comes next follows the same familiar path only, being the *Stanford Encyclopedia of Philosophy*, it is far more extensive than most sources. In fact, the remainder of the entry is over *nineteen thousand* words long and explores scores of specific issues, theories, disputes, and occasional resolutions in the field without, of course, any suggestion of an effort at a definition.

Wikipedia's page tells a similar story though it does have a telling sentence in its rather extensive entry:

> Despite the difficulty in definition, many philosophers believe that there is a broadly shared underlying *intuition* about what consciousness is. [Emphasis added.]

We share this sentiment.

Searches through the writings of individual scholars who study consciousness yielded pretty much the same looseness of meaning. Neither Owen Flanagan's respected *Consciousness Reconsidered* nor David Chalmers' controversial *The Conscious Mind* seriously attempt a definition. In *The Nature of Consciousness: Philosophical Debates*, edited by Ned Block, Flanagan, and Güven Güzeldere, none of the many authors is particularly helpful; ditto with those who contributed to Jonathan Cohen and Jonathan Schooler's edited, interdisciplinary volume *Scientific Approaches to Consciousness.*

It really was becoming amusing. No matter how far afield we went we could find no genuine lexicographic help anywhere. It was not as if authors were all consciously(!) avoiding the issue—though it was pretty clear that more than a few were. And it was not that that the sorts of definitions that were being provided just did not satisfy. In the few cases where someone actually made an attempt to come to grips with the meaning of the term, they usually produced (a) variations of the 'definitions' given by Sutherland and Blackburn, that is, lists of issues that have arisen when contemplating consciousness; or (b) complaints about how no one really has a coherent sense of what consciousness is other than that personal sense that each of us has of whatever the heck it is that each of us certainly has.

John Searle, not surprisingly, has also tackled the definitional problem and, not surprisingly, he feels that he has a way to deal with it. From his book, *The Mystery of Consciousness*:

> One often hears it said that 'consciousness' is frightfully hard to define. But if we are talking about a definition in common sense terms, sufficient to identify the target of the investigation, as opposed to a precise scientific definition of the sort that typically comes at the end of a scientific investigation, then the word does not seem to me hard to define. Here is the definition: Consciousness consists of inner, qualitative, subjective states and processes of sentience or awareness. Consciousness, so defined, begins when we wake in the morning from a dreamless sleep—and continues until we fall asleep again, die, go into a coma or otherwise become 'unconscious'. It includes all of the enormous variety of the awareness that we think of as characteristic of our waking life. It includes everything from feeling a pain, to perceiving objects visually, to states of anxiety and depression, to working out cross word puzzles, playing chess, trying to remember your aunt's phone number, arguing about politics, or to just wishing you were somewhere else. Dreams on this definition are a form of consciousness, though of course they are in many respects quite different from waking consciousness.

This is not bad, not at all. It is vintage Searle with its focus on the empirically demonstrable but when you unpack this paragraph it really turns out to be a compendium of those mental events, states, and experiences that he feels should be tucked under the umbrella term itself—which is pretty much the same place where we arrived when embracing the folk-psychology meaning. Searle, of course, acknowledges this when he tosses in this line: '[W]e are talking about a definition in common sense terms, sufficient to identify the target of the investigation.'

Of course, none of this denotative turmoil is new. George Miller, one of the first to mount a serious research programme in cognitive psychology, fretted about how this favourite topic of his had been abused. He manifested his unhappiness in his usual, elegant way. For him, consciousness had become:

> A word worn smooth by a million tongues. Depending upon the figure of speech chosen it is a state of being, a substance, a process, a place, an epiphenomenon, an emergent aspect of matter, or the only true reality. . . . Perhaps we become confused because whenever we are thinking about consciousness, we are surrounded by it, and can only imagine what consciousness is *not*.

It is worth noting that George Miller wrote this back when psychology was just beginning to emerge from decades of dominance by behaviourism—and, amusingly, penned while he and the unrepentant B. F. Skinner, who regarded such topics as unfit for the attention of serious scientists, were colleagues in the same department at Harvard.

This contorted denotative and connotative mess gains us, we hope, a little sympathy for our bailing out on any serious efforts to 'damn it, define your terms' or to track down the meanings of the many others like *mind, subjective, phenomenal, sentience, awareness*, or *self* and *self-aware*—all of which have, for centuries now, been worked over by many.

Looking back over this material you can see that strong tendency to tie the various meanings to the human experience, to our kinds of subjectivity, our kinds of minds. Essentially all took on the task of defining consciousness beginning with the presumption that it was a mental state primarily experienced by members of our species. While it is true that there were a few efforts such as Donald Griffin's (1992) that did touch upon the possibility of animal minds, or for that matter, plant or protozoan minds, it was rare to find an effort at a definition that was free from species narcissism. And even when the possibility of consciousness in another species was considered plausible, the subjective state was still viewed within the context of, or by comparison with, human experience. Because the focus of the CBC model is on the initial emergence of sentience, when all was said and done, it just seemed wisest to fall back on the folk psychology senses of the many terms used as loose synonyms for consciousness. This, of course, is precisely what virtually all of those working in the many overlapping and interlocking fields of consciousness science have done.

APPENDIX II

Glossary of Technical Terms in the Biological Sciences

As noted in the Preface, in order to understand the many complex functions of cells that are responsible for cellular sentience, we needed to examine the underlying biomolecular mechanisms responsible. To smooth the way for readers lacking a background in cell biology, we have compiled this glossary. The definitions are straightforward and as non-technical as possible. Note that when a term in a definition is in boldface, it means that it has its own entry.

action potentials: The rapid rise and subsequent fall of **voltage** across **biomembranes**.

adenosine diphosphate (ADP): ATP molecule lacking one phosphate group.

adenosine triphosphate (ATP): An energy-carrying molecule found in all living cells.

ADP: See **adenosine diphosphate**.

algae: Photosynthetic unicellular protists.

anaesthetics: Compounds that act on specific, sensitive, cellular targets, especially **biomembranes**. They induce diminished sensitivity (analgesia) and loss of consciousness (anaesthesia) depending on biochemical properties and dosage.

archaea: Prokaryotic, unicellular organisms lacking a nucleus. However, they have numerous eukaryotic features and are considered possibly the most ancient form of life. See **eukaryote** and **prokaryote**.

ATP: See **adenosine triphosphate**.

ATPases: A class of enzymes that catalyse transformation of **ATP** into **adenosine diphosphate** (ADT) for cellular energy deployment.

axons: Projections of neurons that conduct action potentials away from the neuronal cell body.

basal bodies: Protein complexes that function in the assembly of eukaryotic **flagella** and **cilia**.

bioelectric code: Information encoded through bioelectric states of cellular molecules, structures, and processes that are essential for guiding growth, development, morphogenesis, and behaviour of organisms.

bioelectric field: An electric field generated by fluxes of charged ions across **biomembranes**.

biogenesis: Life from life. Generation of living organisms from pre-existing life, particularly of the same type.

biological Maxwell's demons: Cellular systems with information processing capabilities contributing to intracellular structural order. They operate by controlling what is allowed to enter and leave a cell through its membrane. The name comes from the thought experiment proposed by James Clerk Maxwell in 1867 to test the second law of thermodynamics.

biological order: The totality of the ordered cellular structures and processes essential for the living status of cells.

bioluminescence: Internally generated light emission by organisms in which light energy is released by biochemical reactions based on a light-emitting pigment **luciferin** interacting with the oxidative enzyme **luciferase**.

biomembrane: The selectively permeable sheet-like barrier separating the inside contents of cells from the external environment.

biophotons: Photons of light in the ultraviolet and ultra-weak ranges produced by electronically excited biomolecules, especially those associated with the production of **reactive molecular species.**

bioplasma: Networks of electromagnetic fields generated within cells.

biopoiesis: Life from non-life. The concept that life develops from non-living matter and the basis for a widely accepted theory of the origin of life on Earth.

biosemiotics: The integration of biology and semiotics treating all organisms as sign-generating and information-interpreting living systems.

biosphere: The sum of all ecosystems on Earth as they evolved through the functions of **biopoiesis** and **biogenesis.**

calcium waves: Wave-like, localized increases of **cytoplasmic calcium.**

carnivorous plants: Plants that have evolved leaf-traps to hunt, capture, and consume insects and small animals.

cell sentience: Cellular awareness of the external environment and itself.

cellular circadian clocks: Biochemical oscillators within cells that are synchronized with solar time.

cellular communication: Communication within and between cells.

cellular organelles: Membrane-bound compartments within eukaryotic cells. See **organelles.**

cellular respiration: The set of metabolic reactions within cells that allows conversion of chemical energy of organic macromolecules into **ATP** molecules. In eukaryotic cells, cellular respiration is accomplished within their symbiotic (see **symbiosis**) **mitochondria.**

central dogma: The concept that the flow of genetic information is unidirectional from DNA to RNA and proteins. This outdated dogma asserted that once information had gone from DNA into proteins, it could not go back into the genetic code. This unidirectional process is now known to be incorrect.

centrioles: Cylindrical structures composed of **tubulin** proteins organized as nine sets of short **microtubule** triplets. They play a role in organizing the microtubules that serve as the cell's skeletal system.

centrosomes: Organelles that serve as the main structures allowing seeding and **polymerization** of microtubules in eukaryotic cells. Typically, they consist of two **centrioles** surrounded by pericentriolar material.

chemiosmotic theory: A theory based on **ATP** synthesis through proton electrochemical coupling. In this process the electron-transport chain is arranged so that an energy-rich proton gradient is generated across the inner membrane of a mitochondrion, driving ATP synthesis.

cilia: Microscopic, hair-like protrusions on the surfaces of many eukaryotes. They are organized by the cell's basal bodies and composed of **microtubules** and microtubular motor proteins. In some organisms they vibrate and cause changes in the surrounding fluids, in others they enable cellular locomotion.

ciliary rootlet: Intracellular extension of ciliary basal bodies (see **basal body**) providing structural support to a cilium.

ciliates: Protists that are covered by numerous cilia.

coacervate: Aqueous phase rich in biological macromolecules formed as large colloidal particles that precipitate out in aqueous medium. Coacervates are considered the first pre-cells which gradually transformed into living cells.

coevolution: The process where one species evolves in response to the evolutionary changes in another species. The changes themselves typically cause reciprocal adaptation.

cyanobacteria: Ancient photosynthetic bacteria that still thrive in all planetary environments.

cytokinesis: The process of cell division where a 'mother' cell separates into two 'daughter' cells. See **mitosis.**

cytomatrix: The complex cooperative macromolecular relationship between cellular interior and the complex set of interior and exterior water networks.

cytoplasm: All the internal cellular content of a cell, except for the nucleus and vacuoles.

cytoplasmic calcium: Calcium ions freely localized within the **cytoplasm**.

cytoplasmic streaming: Flow of the **cytoplasm** within the cellular interior.

cytoskeleton: The network of protein **polymers** forming filaments and tubules in the **cytoplasm** of many living cells, giving them shape and coherence.

dendrites: Branched tubular extensions of neurons which pick up impulses from neighbouring neurons.

dendritic spines: Protrusions on the **dendrites** of neurons which receive impulses from axons of adjacent neurons.

dimerize: The process in which two molecules combine and create a single **polymer** called a 'dimer'.

electrome: The totality of ionic currents and signals within an individual cell.

electron transport chain: The transfer of electrons from electron donors to electron acceptors enabling their transmembrane transport. The process functions by driving the creation of ATP used by the cell as energy for metabolic processes and cellular functions.

electrophysiology: The physiology of organisms with a focus on bioelectric phenomena.

electrostatics: A subdiscipline of biophysics studying electrical charges of ions and membranes.

embryogenesis: The development of an embryo from a fertilized oocyte (egg cell).

endocytosis: A process of the internalization of a part of the extracellular space into the cell by membrane **invagination**.

endogenous anaesthetics: Anaesthetic compounds synthesized internally in organisms under stress. Many plant species produce their own anaesthetics.

endosomes: Intracellular, membranous compartments generated by the process of **endocytosis.**

endosymbiosis: A form of **symbiosis** where the symbiont lives within the body of a host organism. It is almost certainly the process whereby the first **eukaryotes** evolved.

engram: A unit of cognitive information imprinted in cellular structures.

entropy: Disorder, randomness, and uncertainty. The second law of thermodynamics states that in all closed systems, entropy always increases. Organisms, in their goal for survival, appear to act against entropy to create states of greater order.

enzymes: Proteins that act as catalysts accelerating biological reactions.

epigenesis: Formation of organisms from an original seed, spore, or fertilized egg. Epigenesis replaced the now discredited theory of **preformationism**.

epigenome: The sum of all non-coding DNA/RNA-based hereditary information of an organism. Besides the machinery regulating the packaging of DNA into chromatin and chromosomes, the epigenome includes all cellular structures acting as templates for their own replication. It is one means by which organisms adapt to the environment without requiring a change in the genetic code.

ether: An anaesthetic agent that induces loss of pain sensing (analgesia) and loss of consciousness (anaesthesia).

ethylene: A plant stress hormone that also acts as a general anaesthetic agent.

eukaryotes: Complex cells with a nucleus containing the cell's genome and a large number of intracellular **organelles**. The first eukaryotes are believed to have been formed through **endosymbiosis** where a small **prokaryote** was incorporated by a larger one.

excitability: The capacity of cells to be excited by stimuli. It's a fundamental function for survival of all cellular life and based on processes generated by biomolecular operations on the cell's excitable plasma **membrane**.

exosome: A extracellular **vesicle** released from eukaryotic cells.

exteroception: Sensing of events and stimuli in the external environment. It is an essential feature of the sensory/perceptual processes of all cells and more complex organisms (compare with **interoception**).

flagella: The membranous protrusions from cells used both for motility and as a sensing apparatus.

flagellate: 1. A hair-like protrusion on a cell's membrane that can be rotated or used in a whip-like fashion for propulsion. 2. A group of cells, mostly **protists** and male sex cells, that use **microtubule**-based flagella for movements.

frugivores: Animals that feed primarily on fruits and other plant organs.

gamete: See **sex cells**.

gene: The basic unit of genetic information. The term **gene** was introduced by Danish plant physiologist and geneticist Wilhelm Johannsen in 1909. The concept of a gene is still evolving and has several different meanings.

gene expression: The transfer of information encoded in sequences of **nucleotides** used in the synthesis of proteins, transfer RNAs, small nuclear RNAs, and non-coding RNAs enabling biological actions.

genome: The sum of all protein-coding DNA/RNA-based hereditary information of an organism.

holobiont: A multicellular, eukaryotic organism comprised of both its own eukaryotic cells and its partnering **microbiome**, forming a cohesive, adaptive unit through **symbiosis**.

homeorhesis: The tendency of organisms to return back to their original trajectories. A 'trajectory' can be either movement based or an internal, sentient process. Compare with **homeostasis** in which organism returns to states of stability.

homeostasis: The tendency of organisms to return to their original biological, physiological state. Compare with **homeorhesis** where the tendency is for an organism to return to an original trajectory.

hydrophobic pockets: Protein domains proposed to act as the primary targets of anaesthetics.

hyphae: Long fungal filaments that can branch, fuse, or join together to form mycelia cords (see **cilia**).

info-autopoiesis: The process of the internal self-production of information. All the information that any cell has about events or stimuli in the environment must cross the plasma **membrane** and is ultimately self-interpreted inside the cell.

interfacial water: A unique state of water with high viscosity and ordered structure having separate thermodynamic features.

interoception: The sensory/perceptual processing of internal operations and status of all cells and more complex organisms. Compare with **exteroception**.

invagination: Process of **biomembrane** folding to form a cavity which can be pinched off as a **vesicle**.

ions: Molecules with electrical charges.

ionic wave: The flow of ions along cellular structures.

lichens: Composite organisms based on fungi, algae, or cyanobacteria.

lipid bilayer: The double-layer of **phospholipids** that forms the main structure of a **biomembrane**.

lipid rafts: Domains of **biomembranes** enriched with ordered **sterols**.

luciferase: An oxidative enzyme producing **bioluminescence** from **luciferin**.

luciferin: A biomolecule that emits light in its excited state through enzymatic activity of **luciferase**.

macromolecular crowding: A condition that emerges when biological macromolecules form collectives at very high concentrations.

membrane: See **biomembrane**.

membrane potential: The difference in the electric potential between the inside and outside of the **biomembrane**. The interior is typically negative with respect to the exterior, a feature that generates energy across the membrane, allowing the transmission of molecular signals between the cellular interior and the external environment.

membrane repair: Generally, any of several processes that maintain the structural and functional integrity of **biomembranes**.

microbiome: A community of microorganisms living together with their host organism.

microtubules: Polymers of **tubulin** representing part of the cytoskeleton of eukaryotic cells. Typically, 13 protofilaments join together to form a microtubule.

mimicry: The ability of organisms to imitate features of other organisms for their own advantage and protection.

mitochondria: Symbiotic organelles of most eukaryotic cells responsible for the cellular aerobic respiration that produces **adenosine triphosphate** (ATP). Believed to have originally been a free-living bacterium that became incorporated in a host cell. See **symbiosis**.

mitosis: A process where a single cell divides into two 'daughter' cells with identical genomic complements and provides each with an equal division of the plasma membrane. Mitosis is the typical process of tissue growth and development.

molecular motors: Protein complexes that use **ATP** to fuel their movements.

morphogenesis: The processes that coordinate the origin and development of physical characteristics.

multi-vesicular bodies: A special class of late **endosomes** which internalize **vesicles**.

mycorrhiza fungi: Symbiotic fungi that colonize plant roots. These ancient fungi, which first evolved some 460 million years ago, allowed early plants to live on land.

nano-intentionality: Intentional faculties of biological macromolecules allowing intrinsic goal-directedness of all cells. Nano-intentionality is specific for living cells, and helps explain the difference between computer architecture, AI, and living systems.

nanotube: A microscopic tubular, membranous connection between cells forming **tubular nanotubes** which play a role in communication and the transfer of cellular resources.

N-space: A term derived from mathematics and used to describe the four-dimensional fabric that represents the totality of universal information potentially available to appropriate observers.

N-space Episenome: A multicellular partitioning of **N-space** that structures multicellular **morphogenesis** and the coordinated **embryogenesis**. The N-space Episenome is a heritable, adaptive, information blueprint that guides growth and development from conception forward and is essential for reproduction and development.

nucleotides: Organic molecules consisting of a nucleoside and a phosphate and serving as monomeric units of the nucleic acid polymers—deoxyribonucleic acid (DNA) and ribonucleic acid (RNA). Nucleotides are the basic structural unit of DNA and RNA.

ontogenesis: Development of organisms from oocytes (eggs) fertilized by sperm cells.

organelles: Localized structured compartments of eukaryotic cells each of which performs specific cellular functions.

oxidative phosphorylation: A cell process that harnesses the reduction of oxygen, generating high-energy phosphate bonds to form **adenosine triphosphate** (ATP) which is necessary for metabolism in all organisms.

peripersonal space: The space surrounding an organismal body that is in reach of, or can be reached by, adjacent organisms as an essential component of bodily self-consciousness.

pervasive information field (PIF): A specific sub-domain compartment of universal **N-space** representing the totality of all potential information available to a cell at any moment. This subset of bio-information is required by cells for effective information management.

phospholipids: A class of lipids which self-assemble the lipid bilayers of **biomembranes**.

photosynthesis: The process of converting the light energy of photons into the chemical energy of biological molecules.

piezoelectricity: The accumulation of electric charge in some biomolecules in response to mechanical stress. In cells, the bioelectricity induced by pressure.

PIF: see pervasive information field.

plant sexuality: The sexual reproduction of plants through the fertilization of female sex cells (egg cells) by male sex cells (sperm cells).

pollen tubes: Tip-growing tubular plant cells that act as carriers of male sex cells.

pollinators: Insects or other organisms that transport plant pollen grains from flower to flower, serving plants in their sexual reproduction.

polymerization: The process of **polymer** assembly from associating monomers.

polymers: Biological macromolecules composed of numerous repeating subunits (monomers).

preformationism: The once-popular theory that organisms developed from their own miniature versions. For example, a human sperm cell was thought to have a tiny person inside and often depicted as such. The term, however, is still used in some contexts to refer to functions that are epigenetic (see **epigenesis**.)

prokaryotes: Single-celled organisms, including bacteria and archaea, that, unlike **eukaryotes**, lack a nucleus. Their DNA is distributed loosely throughout the organism's interior.

proprioception: The sense of body location, position, and movement.

proteome: The sum of all cellular proteins.

protists: Single-celled **eukaryotes** such as **protozoa** and simple **algae**.

proto-cells: Non-living, self-organized nano- and micro-sized biochemical structures that formed in the prebiotic 'soup' and are considered the likely progenitors of fully competent cells.

proton-motive force: The biological processes whereby proton gradients operate across membranes driving **ATP** synthesis.

protoplast: A term proposed by Johannes von Hanstein for the entire cell interior, excluding the cell wall and extracellular matrix.

protozoa: A class of single-celled eukaryotic organisms including amoeba and paramecia, among others.

radical pairs of electrons: Correlated electron spins involved in the magneto-perception of birds and a critical element in seasonal migration. Interestingly, they are also an element in xenon-induced general anaesthesia.

reactive molecular species: Reactive and charged biomolecules produced through cellular metabolism and signalling pathways. Note that 'species' here, and in the following entry, refers to a group of radicals, not a genetically related group of organisms.

reactive oxygen species (ROS): An unstable molecule that contains oxygen that easily reacts with others in a cell. Excessive build up is associated with cellular damage, especially to DNA and RNA.

receptors: Biological macromolecules in sense organs. The term has wide, general reference and also denotes structures that are distributed across a cell's biomembranes or an organism's body that are capable of receiving and transmitting signals.

redox: A type of chemical reaction in which the oxidation states of biomolecules change. When electrons are lost, oxidation occurs or is increased; a gain of electrons produces a decrease in oxidation.

redox code: A set of principles governing and maintaining redox **homeostasis**. A redox reaction is a type of chemical interaction that involves a transfer of electrons between two entities.

redox potential: Oxidation/reduction potential measured as the tendency of a chemical species to acquire or lose electrons. Each molecular species has its own intrinsic redox potential. See **reactive molecular species**.

regeneration: The process in which an organism is able to reassemble and reform missing parts or organs.

rhizoplasts: Contractile structures connecting the **basal bodies** of eukaryotic **flagella** and **cilia** with the surface of the nucleus.

root–fungal networks: Huge networks organized by plant roots and **hyphae** of symbiotic fungi which serve in the transport of water and nutrients and as a means of information exchange across whole ecosystems.

ROS: See **reactive oxygen species.**

ROS waves: The flow of **reactive oxygen species** propagating over considerable distances in response to cellular stimulation.

self-organization: In biology, the ability of life processes and biomolecular functions to generate and maintain biological order.

senome: The totality of all bioelectric and sensory activities of **biomembranes** and their associated **organelles** and biomolecules based on the fluxes of all ionic and other cellular charged particles.

senomic field: A bio-electromagnetic field generated by biomembranes and forming a crucial part of the cell's sensory apparatus.

sensors: In biology, sensors are macromolecules, typically protein complexes, which detect some abiotic or biotic parameters and convey sensory information into the cell.

sex cells: Reproductive haploid cells, also known as *gametes*. In multicellular organisms, male sperm cells are smaller and motile, whereas female oocytes are large and non-motile.

spontaneous generation: A theory believed for centuries that organisms regularly arise from non-living matter.

sterols: Organic compounds vital for functions of biomembranes.

symbiosis: A long-term interaction between organisms benefiting all partners involved. Examples include your partnering microbiome or the beneficial relationship between ants and acacia trees. The ants provide the tree with protection against predators and the tree supplies nutrients to the ants. See also, **endosymbiosis.**

synaptic domains: Cell–cell adhesion domains consisting of cell-surface molecules and special types of intercellular bonds critical for cellular communication.

syncope: Loss of consciousness and muscle strength characterized by fast onset, short duration, and spontaneous recovery.

thanatosis: Death-feigning marked by tonic immobility. It is a defensive anti-predatory strategy used by several animal prey species.

tubulin: Globular proteins which **dimerize** and form cytoskeletal **polymers** known as **microtubules.**

tunnelling nanotube: A microscopic cytoplasmic bridge between cells used for cellular communications and exchange of cellular resources. See **nanotube.**

valenced experiences: Context-dependent (subjective) awareness of environmental signals, events, or stimuli. 'Valenced' here simply means that the experiences are marked by either positive or negative feelings.

vesicles: Biomembrane-derived small, intracellular or extracellular compartments enclosed by **lipid bilayers** used to move substances within or outside cells.

viruses: Semi-living infectious agents that reproduce only inside living cells. Viruses infect life forms and provoke immune responses in cells and organisms, but are also crucial partners in cellular adaptation.

vitalism: The view that living organisms are fundamentally different from non-living entities by having some type of special essence, an *élan vital.*

voltage: The difference in electric potential and charges between two points.

Bibliography

Abbas, M., Lipiński, W. P., Wang, J., & Spruijt, E. (2021). Peptide-based coacervates as biomimetic protocells. *Chemical Society Reviews* 50, 3690. https://doi.org/10.1039/d0cs00307g

Abuammar, H., Bhattacharjee, A., Simon-Vecsei, Z., Blastyák, A., Csordás, G., Páli, T., & Juhász, G. (2021). Ion channels and pumps in autophagy: a reciprocal relationship. *Cells* 10, 3537. https://doi.org/10.3390/cells10123537

Adamatzky, A. & Gandia, A. (2022). Fungi anaesthesia. *Scientific Reports* 12, 340. https://doi.org/10.1038/s41598-021-04172-0

Adamski, A. G. (2011). Bioplasma concept of consciousness. *NeuroQuantology* 9, 681–691. https://doi.org/10.14704/nq.2011.9.4.382

Adamski, A. G. (2016). Role of Bose-Einstein condensate and bioplasma in shaping consciousness. *NeuroQuantology* 14, 36–44. https://doi.org/10.14704/NQ.2016.14.1.896

Adamo, G., Fierli, D., Romancino, D. P., Picciotto, S., Barone, M. E., Aranyos, A., et al. (2021). Nanoalgosomes: introducing extracellular vesicles produced by microalgae. *Journal of Extracellular Vesicles* 10, e12081. https://doi.org/10.1002/jev2.12081

Adamo, S. A. (2016.). Consciousness explained or consciousness redefined? *Proceedings of the National Academy of Sciences USA* 113, E3812. https://doi.org/10.1073/pnas.1606942113

Agnati, L. F., Baluška, F., Barlow, P. W., & Guidolin, D. (2009). Mosaic, self-similarity logic and biological attraction principles: three explanatory instruments in biology. *Communicative & Integrative Biology* 2, 552–556. https://doi.org/10.4161/cib.2.6.964

Ainsworth, T. D. & Brown, B. E. (2021). Coral bleaching. *Current Biology* 31, R5–R6. https://doi.org/10.1016/j.cub.2020.10.048

Akıl, C. & Robinson, R. C. (2018). Genomes of Asgard archaea encode profilins that regulate actin. *Nature* 562, 439–443. https://doi.org/10.1038/s41586-018-0548-6

Akıl, C., Kitaoku, Y., Tran, L. T., Liebl, D., Choe, H., Muengsaen, D., et al. (2021). Mythical origins of the actin cytoskeleton. *Current Opinion in Cell Biology* 68, 55–63. https://doi.org/10.1016/j.ceb.2020.08.011

al'Absi, M., Nakajima, M., & Bruehl, S. (2021). Stress and pain: modality-specific opioid mediation of stress-induced analgesia. *Journal of Neural Transmission* 128, 1397–1407. https://doi.org/10.1007/s00702-021-02401-4

Alberts, B., Johnson, A., Lewis, J., Raff, M., Roberts, K., & Walter, P. (2002). The RNA world and the origins of life. In: *Molecular Biology of the Cell*, 4th edition. New York: Garland Science. https://www.ncbi.nlm.nih.gov/books/NBK26876/

Alboni, P. & Alboni, M. (2014). Vasovagal syncope as a manifestation of an evolutionary selected trait. *Journal of Atrial Fibrillation* 7, 1035. https://doi.org/10.4022/jafib.1035

Allen, G. E. (2005). Mechanism, vitalism and organicism in late nineteenth and twentieth-century biology: the importance of historical context. *Studies in History and Philosophy of Science Part C: Studies in History and Philosophy of Biological and Biomedical Sciences* 36, 261–283.

Allen, N. S. & Allen, R. D. (1978). Cytoplasmic streaming in green plants. *Annual Review of Biophysics and Bioengineering* 7, 497–526. https://doi.org/10.1146/annurev.bb.07.060178.002433

Almeida, C. & Amaral, M. D. (2020). A central role of the endoplasmic reticulum in the cell emerges from its functional contact sites with multiple organelles. *Cellular and Molecular Life Sciences* 77, 4729–4745. https://doi.org/10.1007/s00018-020-03523-w

Altmäe, S., Franasiak, J. M., & Mändar, R. (2019). The seminal microbiome in health and disease. *Nature Reviews Urology* 16, 703–721. https://doi.org/10.1038/s41585-019-0250-y

Amponsah, P. S., Yahya, G., Zimmermann, J., Mai, M., Mergel, S., Mühlhaus, T., et al. (2021). Peroxiredoxins couple metabolism and cell division in an ultradian cycle. *Nature Chemical Biology* 17, 477–484. https://doi.org/10.1038/s41589-020-00728-9

Anaya-Prado, R. & Schadegg-Peña, D. (2015). Crawford Williamson Long: the true pioneer of surgical anesthesia. *Journal of Investigative Surgery* 28, 181–187. https://doi.org/10.3109/08941939.2015.1061826

Anders, E. (1989). Pre-biotic organic matter from comets and asteroids. *Nature* 342, 255–257. https://doi.org/10.1038/342255a0

Andersen, S. S. L., Jackson, A. D., & Heimburg, T. (2009). Towards a thermodynamic theory of nerve pulse propagation. *Progress in Neurobiology* 88, 104–113. https://doi.org/10.1016/j.pneurobio.2009.03.002

Andrews, N. W., Almeida, P. E., & Corrotte, M. (2014). Damage control: cellular mechanisms of plasma membrane repair. *Trends in Cell Biology* 24, 734–742. https://doi.org/10.1016/j.tcb.2014.07.008

Anten, N. P. R. & Chen, B. J. W. (2021). Detect thy family: mechanisms, ecology and agricultural aspects of kin recognition in plants. *Plant Cell & Environment* 44, 1059–1071. https://doi.org/10.1111/pce.14011

Aon, M. A. & Cortassa, S. (2015). Function of metabolic and organelle networks in crowded and organized media. *Frontiers in Physiology* 5, 523. https://doi.org/10.3389/fphys.2014.00523

Arkhipova, I. R. & Yushenova, I. A. (2019). Giant transposons in eukaryotes: is bigger better? *Genome Biology and Evolution* 11, 906–918. https://doi.org/10.1093/gbe/evz041

Asakura, M., Matsumura, K., Ishihara, R., & Miyatake, T. (2021). Freezing or death feigning? Beetles selected for long death, feigning showed different tactics against different predators. *Ecology and Evolution* 12, e8533. https://doi.org/10.1002/ece3.8533

Ashe, A., Colot, V., & Oldroyd, B. P. (2021). How does epigenetics influence the course of evolution? *Philosophical Transactions of the Royal Society B* 376, 20200111. https://doi.org/10.1098/rstb.2020.0111

Avidor-Reiss, T., Mazur, M., Fishman, E. L., & Sindhwani, P. (2019). The role of sperm centrioles in human reproduction—the known and the unknown. *Frontiers in Cellular and Developmental Biology* 7, 188. https://doi.org/10.3389/fcell.2019.00188

Avidor-Reiss, T., Achinger, L., & Uzbekov, R. (2022). The centriole's role in miscarriages. *Frontiers in Cellular and Developmental Biology* 10, 864692. https://doi.org/10.3389/fcell.2022.864692

Baggini, J. (2015). *Freedom Regained: The Possibility of Free Will.* Chicago, IL: University of Chicago Press.

Bakillah, A., Al Hejji, F., Almasaud, A., Al Jami, H., Hawwari, A., Qarni, A. A., et al. (2022). Lipid raft integrity and cellular cholesterol homeostasis are critical for SARS-CoV-2 entry into cells. *Nutrients* 14, 3417. https://doi.org/10.3390/nu14163417

Ball, P. (2008). Water as an active constituent in cell biology. *Chemical Reviews* 108, 74–108. https://doi.org/10.1021/cr068037a

Ball, P. (2016). The problems of biological information. *Philosophical Transactions of the Royal Society A: Mathematical, Physical and Engineering Sciences* 374, 20150072. https://doi.org/10.1098/rsta.2015.0072

Ball, P. (2017). Water is an active matrix of life for cell and molecular biology. *Proceedings of the National Academy of Sciences USA* 114, 13327–13335. https://doi.org/10.1073/pnas.1703781114

Baluška, F. (2009). Cell-cell channels, viruses, and evolution: via infection, parasitism, and symbiosis toward higher levels of biological complexity. *Annnals of the New York Academy of Sciences* 1178, 106–119. https://doi.org/10.1111/j.1749-6632.2009.04995.x

Baluška, F. (2016). Should fish feel pain? A plant perspective. *Animal Sentience* 3, 16.

Baluška, F., Hlavacka, A., Volkmann, D., & Menzel, D. (2004). Getting connected: actin-based cell-to-cell channel in plants and animals. *Trends in Cell Biology* 14, 404–408. https://doi.org/10.1016/j.tcb.2004.07.001

Baluška, F. & Levin, M. (2016). On having no head: cognition throughout biological systems. *Frontiers of Psychology* 7, 902. https://doi.org/10.3389/fpsyg.2016.00902

Baluška, F. & Lyons, S. (2018a). Energide-cell body as smallest unit of eukaryotic life. *Annals of Botany* 122, 741–745. https://doi.org/10.1093/aob/mcy022

Baluška, F. & Lyons, S. (2018b). Symbiotic origin of eukaryotic nucleus—from cell body to neo-energide. In: Sahi, V. & Baluška, F. (Eds.), *Concepts in Cell Biology—History and Evolution*, pp. 39–66. Cham: Springer.

Baluška, F. & Lyons, S. (2021). Archaeal origins of eukaryotic cell and nucleus. *Biosystems* 203, 104375. https://doi.org/10.1016/j.biosystems.2021.104375

Baluška, F. & Mancuso, S. (2009a). Deep evolutionary origins of neurobiology: turning the essence of 'neural' upside-down. *Communicative & Integrative Biology* 2, 60–65. https://doi.org/10.4161/cib.2.1.7620

Baluška, F. & Mancuso, S. (2009b). Plants and animals: convergent evolution in action? In: Baluška, F. (Ed.), *Plant-Environment Interactions*, pp. 285–301. Springer-Verlag: Berlin Heidelberg. https://doi.org/10.1007/978-3-540-89230-4_15

Baluška, F. & Mancuso, S. (2013). Ion channels in plants: from bioelectricity, via signaling, to behavioral actions. *Plant Signaling & Behavior* 8, e23009. https://doi.org/10.4161/psb.23009

Baluška, F. & Mancuso, S. (2014). Synaptic view of eukaryotic cell. *International Journal of General Systems* 43, 740–756. https://doi.org/10.1080/03081079.2014.920999

Baluška, F. & Mancuso, S. (2019). Actin cytoskeleton and action potentials: forgotten connections. In: Baluška, F. & Sahi, V. (Eds.), *The Cytoskeleton*, pp. 63–83. Cham: Springer. https://doi.org/10.1007/978-3-030-33528-1_5

Baluška, F. & Mancuso, S. (2020). Plants, climate and humans. Plant intelligence changes everything. *EMBO Reports* 21, e50109. https://doi.org/10.15252/embr.202050109

Baluška, F. & Mancuso, S., (2021). Individuality, self and sociality of vascular plants. *Philosophical Transaction of the Royal Society London B Biological Sciences* 376, 20190760. https://doi.org/10.1098/rstb.2019.0760

Baluška, F., Mancuso, S., & Van Volkenburgh, E. (2021a). Barbar' Pickard—queen of plant electrophysiology. *Plant Signaling & Behavior* 16, 1911400. https://doi.org/10.1080/15592324.2021.1911400

Baluška, F. & Miller, W. B., Jr. (2018). Senomic view of the cell: senome versus genome. *Communicative & Integrative Biology* 11, 1–9. https://doi.org/10.1080/19420889.2018.1489184

Baluška, F., Miller, W. B., Jr., & Reber, A. S. (2021b). Biomolecular basis of cellular consciousness via subcellular nano-brains. *International Journal of Molecular Sciences* 22, 2545. https://doi.org/10.3390/ijms22052545

Baluška, F., Miller, W. B., Jr., & Reber, A. S. (2023in press). Cellular basis of cognition—from protozoa and fungi up to plants and root-fungal networks. In: Corning, P. A., Kauffman, S. A., Noble, D., Shapiro, J. A., Vane-Wright, R. I., & Pross, A. (Eds.), *Evolution 'On Purpose': Teleonomy in Living Systems*. Cambridge, MA: MIT Press.

Baluška, F., Miller, W. B., Jr., & Reber, A. S. (2022a). Cellular and evolutionary perspectives on organismal cognition: from unicellular to multicellular organisms. *Biological Journal of the Linnean Society* 137, blac005. https://doi.org/10.1093/biolinnean/blac005

Baluška, F. & Reber, A. (2019). Sentience and consciousness in single cells: how the first minds emerged in unicellular species. *BioEssays* 41, e1800229. https://doi.org/10.1002/bies.201800229

Baluška, F. & Reber, A. (2021a). The biomolecular basis for plant and animal sentience: senomic and ephaptic principles of cellular consciousness. *Journal of Consciousness Studies* 28, 31–49. https://www.ingentaconnect.com/contentone/imp/jcs/2021/00000028/f0020001/art00003

Baluška, F. & Reber, A. S. (2021b). CBC-clock theory of life—integration of cellular circadian clocks and cellular sentience is essential for cognitive basis of life. *BioEssays* 43, e2100121. https://doi.org/10.1002/bies.202100121

Baluška, F. & Reber, A. S. (2022). Cellular and organismal agency—not based on genes: a comment on Baverstock. *Progress in Biophysics and Molecular Biology* 167, 161–162. https://doi.org/10.1016/j.pbiomolbio.2021.11.001

Baluška, F., Reber, A. S., & Miller, W. B. (2022b). Cellular sentience and cognition as the primary source of biological order and evolutionary creativity. *Biosystems* 218, 104694. https://doi.org/10.1016/j.biosystems.2022.104694

Baluška, F., Volkmann, D., & Barlow, P. W. (1997). Nuclear components with microtubule-organizing properties in multicellular eukaryotes: functional and evolutionary considerations. *International Reviews of Cytology* 175, 91–135. https://doi.org/10.1016/s0074-7696(08)62126-9

Baluška, F., Volkmann, D., & Barlow, P. W. (2004a). Eukaryotic cells and their cell bodies: cell theory revised. *Annals of Botany* 122, 741–745. https://doi.org/10.1093/aob/mch109

Baluška, F., Volkmann, D., & Barlow, P. W. (2004b). Cell bodies in a cage. *Nature* 428, 371. https://doi.org/10.1038/428371a

Baluška, F., Volkmann, D., & Barlow, P. W. (2006). Cell-cell channels and their implications for cell theory. In: Baluška, F., Volkmann, D., & Barlow, P. W. (Eds.), *Cell-Cell Channels*, pp. 1–18. Berlin: Landes Bioscience, Springer.

Baluška, F., Volkmann, D., Menzel, D., & Barlow, P. W. (2012). Strasburger's legacy to mitosis and cytokinesis and its relevance for the cell theory. *Protoplasma* 249, 1151–1162. https://doi.org/10.1007/s00709-012-0404-8

Baluška, F. & Wan, Y. L. (2012). Physical control over endocytosis. In: PlantsŠamaj, J. (Ed.), *Endocytosis in Plants*, pp. 123–149. Berlin: Springer Verlag.

Baluška, F. & Yokawa, K. (2021). Anaesthetics and plants: from sensory systems to cognition-based adaptive behaviour. *Protoplasma* 258, 449–454. https://doi.org/10.1007/s00709-020-01594-x

Baluška, F., Yokawa, K., Mancuso, S., & Baverstock, K. (2016). Understanding of anesthesia—why consciousness is essential for life and not based on genes. *Communictive & Integrative Biology* 9, e1238118. https://doi.org/10.1080/19420889.2016.1238118

Barker, S. A. (2018). N, N-dimethyltryptamine (DMT), an endogenous hallucinogen: past, present, and future research to determine its role and function. *Frontiers in Neuroscience* 12, 536. https://doi.org/10.3389/fnins.2018.00536

Barron, A. B., & Klein, A. (2016). What insects can tell us about the origins of consciousness. *Proceedings of the National Academy of Science USA* 113, 4900–4908. https://doi.org/10.1073/pnas.1520084113

Bateson, G. (1973). *Steps to an Ecology of Mind*. St Albans: Paladin, Frogmore.

Bathia, J., Schröder, K., Fraune, S., Lachnit, T., Rosenstiel, P., & Bosch, T. C. G. (2022). Symbiotic algae of *Hydra viridissima* play a key role in maintaining homeostatic bacterial colonization. *Frontiers in Microbiology* 13, 869666. https://doi.org/10.3389/fmicb.2022.869666

Bauer, E. (1935). *Theoretical Biology*. Budapest: Akadémiai Kiadó (published 1984).

Bayliss, R. & Piquet, V. (2018). Masters of manipulation: viral modulation of the immunological synapse. *Cellular Microbiology* 20, e12944. https://doi.org/10.1111/cmi.12944

Beagle, S. D. & Lockless, S. W. (2015). Microbiology: electrical signaling goes bacterial. *Nature* 527, 44–45. https://doi.org/10.1038/nature15641

Becker, B. (2013). Snow ball earth and the split of Streptophyta and Chlorophyta. *Trends in Plant Science* 18, 180–183. https://doi.org/10.1016/j.tplants.2012.09.010

Behrens, H. M., Weisenseel, M. H., & Sievers, A. (1982). Rapid changes in the pattern of electric current around the root tip of *Lepidium sativu.'* following gravistimulation. *Plant Physiology* 70, 1079–1083. https://doi.org/10.1104/pp.70.4.1079

Beilby, M. J. (2007). Action potential in charophytes. *International Reviews of Cytology* 257, 43–82. https://doi.org/10.1016/S0074-7696(07)57002-6

Beilby, M. J. & Al Khazaaly, S. (2016). Re-modeling Chara action potential: I. From Thiel model of Ca^{2+} transient to action potential form. *AIMS Biophysics* 3, 431–449. https://doi.org/10.3934/biophy.2016.3.431

Ben-Amotz, B. (2022). Electric buzz in a glass of pure water. *Science* 376, 800–8001. https://doi.org/10.1126/science.abo3398

Bennetzen, J. L. & Wang, H. (2014). The contributions of transposable elements to the structure, function, and evolution of plant genomes. *Annual Reviews of Plant Biology* 65, 505–530. https://doi.org/10.1146/annurev-arplant-050213-035811

Bentley, B. L. (1977). Extrafloral nectaries and protection by pugnacious bodyguards. *Annual Review of Ecology and Systematics* 8, 407–27. https://doi.org/10.1146/annurev.es.08.110177.002203

Bentley, M. A., Yates, C. A., Hein, J., Preston, G. M., & Foster, K. R. (2022). Pleiotropic constraints promote the evolution of cooperation in cellular groups. *PloS Biology*, 20, e3001626. https://doi.org/10.1371/journal.pbio.3001626

Bentrup, F. W. (1969). Membrane potential measurements on fertilized *Fucus* eggs. *Naturwissenschaften* 56, 331–332. https://doi.org/10.1007/BF00602184

Berg, H. C. & Brown, D. A. (1972). Chemotaxis in *Escherichia coli* analysed by three-dimensional tracking. *Nature* 239, 500–504. https://doi.org/10.1038/239500a0

Bernard, C. (1878). *Leçonssur les phénomènes de la vie communs aux animauxet aux végétaux. Lectures on Phenomena of Life Common to Animals and Plants.* Paris: Ballliere, and Son

Betts, H. C., Puttick, M. N., Clark, J. W., Williams, T. A., Donoghue, P. C., & Pisani, D. (2018). Integrated genomic and fossil evidence illuminates life's early evolution and eukaryote origin. *Nature Ecology & Evolution* 2, 1556–1562. https://doi.org/10.1038/s41559-018-0644-x

Bezanilla, F. (2018). Gating currents. *Journal of General Physiology* 150, 911–932. https://doi.org/10.1085/jgp.201812090

Bhagooli, R. & Hidaka, M. (2004). Release of zooxanthellae with intact photosynthetic activity by the coral *Galaxea fascicularis* in response to high temperature stress. *Marine Biology* 145, 329–337. https://doi.org/10.1007/s00227-004-1309-7

Bhar, A., Chakraborty, A., & Roy, A. (2021). Plant responses to biotic stress: old memories matter. *Plants (Basel)* 11, 84. https://doi.org/10.3390/plants11010084

Biedrzycki, M. L., Jilany, T. A., Dudley, S. A., & Bais, H. P. (2010). Root exudates mediate kin recognition in plants. *Communicative & Integrative Biology* 3, 28–35. https://doi.org/10.4161/cib.3.1.10118

Bigelow, H. J. (1846). Insensibility during surgical operations produced by inhalation. *Boston Medical and Surgical Journal* 35, 309–317.

Binder, P. M. & Danchin, A. (2011). Life's demons: information and order in biology. What subcellular machines gather and process the information necessary to sustain life? *EMBO Reports* 12, 495–499. https://doi.org/10.1038/embor.2011.83

Birch, J. (2017). Animal sentience and the precautionary principle. *Animal Sentience* 16, 1. https://doi.org/10.51291/2377-7478.1200

Birch, J., Schnell, A. K., & Clayton, N. S. (2020). Dimensions of animal consciousness. *Trends in Cognitive Sciences* 24, 789–801. https://doi.org/10.1016/j.tics.2020.07.007

Birch, J., Burn, C., Schnell, A., Browning, H., & Crump, A. (2021). *Review of the Evidence of Sentience in Cephalopod Mollusks and Decapod Crustaceans*. Report prepared for the London School of Economics.

Blaby, I. K., Blaby-Haas, C. E., Tourasse, N., Hom, E. F. Y., Lopez, D., et al. (2014). The *Chlamydomonas* genome project: a decade on. *Trends in Plant Science* 19, 672–680. https://doi.org/10.1016/j.tplants.2014.05.008

Black, S. G., Arnaud, F., Palmarini, M., & Spencer, T. E. (2010). Endogenous retroviruses in trophoblast differentiation and placental development. *American Journal of Reproductive Immunology* 64, 255–264. https://doi.org/10.1111/j.1600-0897.2010.00860.x

Blackall, L. L., Wilson, B., & van Oppen, M. J. H. (2015). Coral—the world's most diverse symbiotic ecosystem. *Molecular Ecology* 24, 5330–5347. https://doi.org/10.1111/mec.13400

Boeke, J. D. & Stoye, J. P. (1997). Retrotransposons, endogenous retroviruses, and the evolution of retroelements. In: Coffin, J. M., Hughes, S. H., & Varmus, H. E. (Eds.), *Retroviruses*, pp. 343–435. Cold Spring Harbor, New York: Cold Spring Harbor Laboratory Press.

Boël, G., Danot, O., de Lorenzo, V., & Danchin, A. (2019). Omnipresent Maxwell's demons orchestrate information management in living cells. *Microbial Biotechnology* 12, 210–242. https://doi.org/10.1111/1751-7915.13378

Bogdanova, O. V., Bogdanov, V. B., Dureux, A., Farnè, A., & Hadj-Bouziane, F. (2021). The peripersonal space in a social world. *Cortex* 142, 28–46. https://doi.org/10.1016/j.cortex.2021.05.005

Boilard, A., Dubé, C. E., Gruet, C., Mercière, A., Hernandez-Agreda, A., & Derome, N. (2020). Defining coral bleaching as a microbial dysbiosis within the coral holobiont. *Microorganisms* 8, 1682. https://doi.org/10.3390/microorganisms8111682

Boisseau, R. P., Vogel, D., & Dussutour, A. (2016). Habituation in non-neural organisms: evidence from slime moulds. *Proceedings of the Royal Society B: Biological Sciences* 283, 20160446. http://dx.doi.org/10.1098/rspb.2016.0446

Boland, F. K. (1923). Crawford Williamson Long and the discovery of anesthesia. *The Georgia Historical Quarterly* 7, 135–154.

Bonner, J. T. (2009). *The Social Amoebae: The Biology of Cellular Slime Molds*. Princeton, NJ: Princeton University Press.

Bonns, W. W. (1918). Etherization of tissues and its effect on enzyme activity. *Annals of the Missouri Botanical Garden* 5, 225–299.

Booker, R. D. & Sum, A. K. (2013). Biophysical changes induced by xenon on phospholipid bilayers. *Biochimica et Biophysica Acta (BBA)—Biomembranes* 1828, 1347–1356. https://doi.org/10.1016/j.bbamem.2013.01.016

Borrel, G., Brugère, J. F., Gribaldo, S., Schmitz, R. A., & Moissl-Eichinger, C. (2020). The host-associated archaeome. *Nature Reviews Microbiology* 18, 622–636. https://doi.org/10.1038/s41579-020-0407-y

Bouteau, F., Grésillon, E., Chartier, D., Arbelet-Bonnin, D., Kawano, T., Baluška, F., et al. (2021). Our sisters the plants? Notes from phylogenetics and botany on plant kinship blindness. *Plant Signaling & Behavior* 16, 2004769. https://doi.org/10.1080/15592324.2021.2004769

Bowler, P. J. (2003). *Evolution: The History of an Idea*. Oakland, CA: University of California Press.

Boyd, J. S., Cheng, R. R., Paddock, M. L., Sancar, C., Morcos, F., & Golden, S. S. (2016). A combined computational and genetic approach uncovers network interactions of the cyanobacterial circadian clock. *Journal of Bacteriology* 198, 2439–2447. https://doi.org/10.1128/JB.00235-16

Boyd, M., Rosenzweig, F., & Herron, M. D. (2018). Analysis of motility in multicellular *Chlamydomonas reinhardtii* evolved under predation. *PLoS One* 13, e0192184. https://doi.org/10.1371/journal.pone.0192184

Boyko, A. & Kovalchuk, I. (2010). Transgenerational response to stress in *Arabidopsis thaliana. Plant Signaling & Behavior* 5, 995–998. https://doi.org/10.1371/journal.pone.0009514

Brady, A., Quiles-Puchalt, N., Del Sol, F. G., Zamora-Caballero, S., Felipe-Ruíz, A., Val-Calvo, J., et al. (2021). The arbitrium system controls prophage induction. *Current Biology* 31, 5037–5045. https://doi.org/10.1016/j.cub.2021.08.072

Bray, D. (2009). *Wetware: A Computer in Every Living Cell.* New York: Elsevier. https://doi.org/10.1016/j.cell.2009.08.024

Brillouin, L. (1951). Maxwell's demon cannot operate: information and entropy. I. *Journal of Applied Physics* 22, 334–337.

Broom, D. M. (2014). *Sentience and Animal Welfare*. Oxfordshire: CAB International.

Brukman, N. G., Li, X., & Podbilewicz, B. (2022). Fusexins, HAP2/GCS1 and evolution of gamete fusion. *Frontiers of Cellular and Developmental Biology* 9, 8240724. https://doi.org/10.3389/fcell.2021.824024

Brunet, T. & Arendt, D. (2016). From damage response to action potentials: early evolution of neural and contractile modules in stem eukaryotes. *Philosophical Transactions of the Royal Society London B Biological Sciences* 371, 20150043. https://doi.org/10.1098/rstb.2015.00439-081629

Bruni, G. N., & Kralj, J. M. (2020). Membrane voltage dysregulation driven by metabolic dysfunction underlies bactericidal activity of aminoglycosides. *eLife*, 9, e58706.

Burki, F., Sandin, M. M., & Jamy, M. (2021). Diversity and ecology of protists revealed by metabarcoding. *Current Biology* 31, R1267–R1280. https://doi.org/10.1016/j.cub.2021.07.066

Burley, S. K. & Kamada, K. (2002). Transcription factor complexes. *Current Opinion in Structural Biology* 12, 225–230. https://doi.org/10.1016/s0959-440x(02)00314-7

Bye, J., Murray, K., & Curtis, R. (2021). ATP and tri-polyphosphate (TPP) suppress protein aggregate growth by a supercharging mechanism. *Biomedicines* 9, 1646. https://doi.org/10.3390/biomedicines9111646

Bywater, R. P. & Conde-Frieboes, K. (2005). Did life begin on the beach? *Astrobiology* 5, 568–574.

Buzás, E. I. (2023). The roles of extracellular vesicles in the immune system. *Nature Reviews Immunology* 23, 236–250. https://doi.org/10.1038/s41577-022-00763-8

Buzsáki, G., Anastassiou, C. A., & Koch, C. (2012). The origin of extracellular fields and currents -EEG, ECoG, LFP and spikes. *Nature Reviews Neurosciences* 13, 407–420. https://doi.org/10.1038/nrn3241

Byrnes, W. M. (2020). E. E. Just's broad, yet hidden, influence on modern cell and developmental biology. *Molecular Reproduction & Development* 87, 380–391. https://doi.org/10.1002/mrd.23270

Cai, Q., Qiao, L., Wang, M., He, B., Lin, F. M., Palmquist, J., et al. (2018). Plants send small RNAs in extracellular vesicles to fungal pathogen to silence virulence genes. *Science* 360, 1126–1129. https://doi.org/10.1126/science.aar4142

Cai, Q., He, B., Wang, S., Fletcher, S., Niu, D., Mitter, N., et al. (2021). Message in a bubble: shuttling small RNAs and proteins between cells and interacting organisms using extracellular vesicles *Annual Reviews of Plant Biology* 72, 497–524. https://doi.org/10.1146/annurev-arplant-081720-010616

Cairns-Smith, A. G. (1971). *The Life Puzzle: On Crystals and Organisms and On the Possibility of a Crystal as an Ancestor*. Toronto: University of Toronto Press.

Callistri, A., Reale, A., Palù, G., & Parolin, C. (2022). Why cells and viruses cannot survive without an ESCRT. *Cells* 10, 483. https://doi.org/10.3390/cells10030483

Camacho, M. P. (2019). The central dogma is empirically inadequate … no matter how we slice it. *Philosophy, Theory, and Practice in Biology* 11, 6. https://doi.org/10.3998/ptpbio.16039257.0011.006

Cameron, L. P. & Olson, D. E. (2018). Dark classics in chemical neuroscience: N, N-dimethyltryptamine (DMT). *ACS Chemical Neuroscience* 9, 2344–2357. https://doi.org/10.1021/acschemneuro.8b00101

Campagna, J. A., Miller, K. W., & Forman, S. A. (2003). Mechanisms of actions of inhaled anesthetics. *New England Journal of Medicine* 348, 2110–2124. https://doi.org/10.1056/NEJMra021261

Carbonara, K., Andonovski, M., & Coorssen, J. R. (2021). Proteomes are of proteoforms: embracing the complexity. *Proteomes* 9, 38. https://doi.org/10.3390/proteomes9030038

Cárdenas-García, J. F. (2020). The process of info-autopoiesis—the source of all information. *Biosemiotics* 13, 199–221. https://doi.org/10.1007/s12304-020-09384-x

Carlesso, D., Smargiassi, S., Pasquini, E., Bertelli, G., & Baracchi, D. (2021). Nectar non-protein amino acids (NPAAs) do not change nectar palatability but enhance learning and memory in honey bees. *Scientific Reports* 11, 11721. https://doi.org/10.1038/s41598-021-90895-z

Carlton, J. G. & Martin-Serrano, J. (2007). Parallels between cytokinesis and retroviral budding: a role for the ESCRT machinery. *Science* 316, 1908–1912. https://doi.org/10.1126/science.1143422

Cartaxana, P., Rey, F., LeKieffre, C., Lopes, D., Hubas, C., Spangenberg, J. E., et al. (2021). Photosynthesis from stolen chloroplasts can support sea slug reproductive fitness. *Proceedings of Royal Society London B—Biological Sciences* 288, 20211779. https://doi.org/10.1098/rspb.2021.1779

Cartwright, J. H., Giannerini, S., & González, D. L. (2016). DNA as information: at the crossroads between biology, mathematics, physics and chemistry. *Philosophical Transactions of the Royal Society A: Mathematical, Physical and Engineering Sciences* 374, 20150071. https://doi.org/10.1098/rsta.2015.0071

Casadesus, J. & D'Ari, R. (2002). Memory in bacteria and phage. *Bioessays* 24, 512–518. https://doi.org/10.1002/bies.10102

Castaneda, N., Feuillie, C. Molinari, M., & Kang, E. H. (2021). Actin bundle nanomechanics and organization are modulated by macromolecular crowding and electrostatic interactions. *Frontiers in Molecular Biosciences* 8, 760950. https://doi.org/10.3389/fmolb.2021.760950

Cervera, J., Meseguer, S., & Mafe, S. (2016). The interplay between genetic and bioelectrical signaling permits a spatial regionalisation of membrane potentials in model multicellular ensembles. *Scientific Reports* 6, 35201. https://doi.org/10.1038/srep35201

Chalmers, D. J. (1996). *The Conscious Mind.* Oxford: Oxford University Press.

Chalmers, D. J. (2022). *Reality+: Virtual Worlds and the Problems of Philosophy.* New York: Norton.

Chandramouly, G., Zhao, J., McDevitt, S., Rusanov, T., Hoang, T., Borisonnik, N., et al. (2021). Polθ reverse transcribes RNA and promotes RNA-templated DNA repair. *Science Advances* 7, eabf1771. https://doi.org/10.1126/sciadv.abf1771

Chapman, M. J. & Margulis, L. (1998). Morphogenesis by symbiogenesis. *International Microbiology* 1, 319–326. PMID: 10943381.

Charles, A. C., Kodali, S. K., & Tyndale, R. F. (1996). Intercellular calcium waves in neurons. *Molecular and Cellular Neuroscience* 7, 337–353. https://doi.org/10.1006/mcne.1996.025

Charras, G. & Lenz, M. (2022). Biochemical timer phases, condensates in and out. *Nature* 609, 469–470. https://doi.org/10.1038/d41586-022-01794-w

Chen, C. G., Nardi, A. N., Amadei, A., & D'Abramo, M. (2022). Theoretical modeling of redox potentials of biomolecules. *Molecules* 27, 1077. https://doi.org/10.3390/molecules27031077

Cheng, K. (2022). Bodily memory in slime mold. *Learning & Behavior* 50, 203–204. https://doi.org/10.3758/s13420-022-00516-6

Chierici, F., Dogariu, A., & Tuszynski, J. A. (2022). Computational investigation of the ordered water system around microtubules: implications for protein interactions. *Frontiers in Molecular Biosciences* 9, 884043. https://doi.org/10.3389/fmolb.2022.884043

Chittka, L. & Wilson, C. (2019). Expanding consciousness. *American Scientist* 107, 364. https://doi.org/10.1511/2019.107.6.364

Chiu, L. & Gilbert, S. F. (2015). The birth of the holobiont: multi-species birthing through mutual scaffolding and niche construction. *Biosemiotics* 8, 191–210. https://works.swarthmore.edu/fac-biology/452

Cho, I. & Blaser, M. J. (2012). The human microbiome: at the interface of health and disease. *Nature Reviews Genetics* 13, 260–270. https://doi.org/10.1038/nrg3182

Choi, J. M., Holehouse, A. S., & Pappu, R. V. (2020). Physical principles underlying the complex biology of intracellular phase transitions. *Annual Reviews of Biophysics* 49, 107–133. https://doi.org/10.1146/annurev-biophys-121219-081629

Choi, W. G., Miller, G., Wallace, I., Harper, J., Mittler, R., & Gilroy, S. (2017). Orchestrating rapid long-distance signaling in plants with Ca^{2+}, ROS and electrical signals. *Plant Journal* 90, 698–707. https://doi.org/10.1111/tpj.13492

Choi, W. G., Toyota, M., Kim, S. H., Hilleary, R., & Gilroy, S. (2014). Salt stress-induced Ca^{2+} waves are associated with rapid, long-distance root-to-shoot signaling in plants. *Proceedings of the National Academy of Sciences USA* 111, 6497–6502. https://doi.org/10.1073/pnas.1319955111

Chu, X. Y., Xu, Y. X., Tong, Y. Y., Wang, G., & Zhang, H. Y. (2022). The legend of ATP: from origin of life to precision medicine. *Metabolites* 12, 461. https://doi.org/10.3390/metabo12050461

Claessen, D., Rozen, D. E., Kuipers, O. P., Søgaard-Andersen, L., & van Wezel, G. P. (2014). Bacterial solutions to multicellularity: a tale of biofilms, filaments and fruiting bodies. *Nature Reviews Microbiology* 12, 115–124. https://doi.org/10.1038/nrmicro3178

Clarke, D., Morley, E., & Robert, D. (2017). The bee, the flower, and the electric field: electric ecology and aerial electroreception. *Journal of Comparative Physiology A* 203, 737–748. https://doi.org/10.1007/s00359-017-1176-6

Clarke, D., Whitney, H., Sutton, G., & Robert, D. (2013). Detection and learning of floral electric fields by bumblebees. *Science* 340, 66–69. https://doi.org/10.1126/science.1230883

Claverie, J. M., Ogata, H., Audic, S., Abergel, C., Suhre, K., & Fournier, P. E. (2006). Mimivirus and the emerging concept of 'giant' virus. *Virus Research* 117(1), 133–144.

Clegg, J. S. (1984a). Intracellular water and the cytomatrix: some methods of study and current views. *Journal of Cell Biology* 99, 167–171. https://doi.org/10.1083/jcb.99.1.167s

Clegg, J. S. (1984b). Properties and metabolism of the aqueous cytoplasm and its boundaries. *American Journal of Physiology* 246, 133–151. https://doi.org/10.1152/ajpregu.1984.246.2.R133

Clegg, J. S. (1992). Cellular infrastructure and metabolic organization. *Current Topics in Cellular Regulation* 33, 3–14. https://doi.org/10.1016/b978-0-12-152833-1.50006-x

Clegg, J. S. & Wheatley, D. N. (1991). Intracellular organization: evolutionary origins and possible consequences to metabolic rate control in vertebrates. *American Zoologist* 31, 504–513. https://doi.org/10.1093/icb/31.3.504

Clemens, W. F. (1848). Sur l'éthérisation des plantes. *Bulletin de la Société Vaudoise des Sciences Naturelles* 2, 289–295.

Coleman, L. S. (2012). 30 years lost in anesthesia theory. *Cardiovascular & Hematological Agents in Medicinal Chemistry* 10, 31–49. https://doi.org/10.2174/187152512799201118

Cohen, S., Valm, A. M., & Lippincott-Schwartz, J. (2018). Interacting organelles. *Current Opinion in Cell Biology* 53, 84–91. https://doi.org/10.1016/j.ceb.2018.06.003

Colizzi, E. S., Vroomans, R. M., & Merks, R. M. (2020). Evolution of multicellularity by collective integration of spatial information. *eLife* 9, e56349. https://doi.org/10.7554/eLife.56349

Collings, D. A., White, R. G., & Overall, R. L. (1992). Ionic current changes associated with the gravity-induced bending response in roots of *Zea may' Plant Physiology* 100, 1417–1426. https://doi.org/10.1104/pp.100.3.1417

Collins, H. (2018). *Artifictional Intelligence: Against Humanity's Surrender to Computers*. Hoboken, NJ: Wiley.

Collins, H. (2022). The unexplicated and its consequences. In: Reber, A. S. & Allen, R. (Eds.), *The Cognitive Unconscious: The First Half-Century*, pp. 288–305. New York: Oxford University Press.

Collins, M.-P. & Forgac, M. (2020). Regulation and function of V-ATPases in physiology and disease. *Biochimica et Biophysica Acta—Biomembranes* 1862, 1833410. https://doi.org/10.1016/j.bbamem.2020.183341

Colman, D. R., Poudel, S., Stamps, B. W., Boyd, E. S., & Spear, J. R. (2017). The deep, hot biosphere: twenty-five years of retrospection. *Proceedings of the National Academy of Sciences USA* 114, 6895–6903. https://doi.org/10.1073/pnas.1701266114

Corcoran, L., Roche, M., & Finn, D. P. (2015). The role of the brain's endocannabinoid system in pain and its modulation by stress. *International Review of Neurobiology* 125, 203–255. https://doi.org/10.1016/bs.irn.2015.10.003

Corsi, R. L., Siegel, J., Karamalegos, A., Simon, H., & Morrison, G. C. (2007). Personal reactive clouds: introducing the concept of near-head chemistry. *Atmospheric Environment* 41, 3161–3165. https://doi.org/10.1016/j.atmosenv.2006.07.054

Cortese-Krott, M. M., Koning, A., Kuhnle, G. G. C., Nagy, P., Bianco, C. L., Pasch, A., et al. (2017). The reactive species interactome: evolutionary emergence, biological significance, and opportunities for redox metabolomics and personalized medicine. *Antioxidant & Redox Signaling* 27, 684–712. https://doi.org/10.1089/ars.2017.7083

Cosetta, C. M., Kfoury, N., Robbat, A., & Wolfe, B. E. (2020). Fungal volatiles mediate cheese rind microbiome assembly. *Environmental Microbiology* 22, 4745–4760. https://doi.org/10.1111/1462-2920.15223

Craddock, T. J. A., Kurian, P., Preto, J., Sahu, K., Hameroff, S. R., Klobukowski, M., et al. (2017). Anesthetic alterations of collective terahertz oscillations in tubulin correlate with clinical potency: implications for anesthetic action and post-operative cognitive dysfunction. *Scientific Reports* 7, 9877. https://doi.org/10.1038/s41598-017-09992-7

Crick, F. (1981). *Life Itself*. New York: Simon and Schuster.

Crick, F. (1970). Central dogma of molecular biology. *Nature* 227, 561–563. https://doi.org/10.1038/227561a0

Crump, A., Browning, H., Schnell, A., Burn, C., & Birch, J. (2022). Sentience in decapod crustaceans: a general framework and review of the evidence. *Animal Sentience* 32, 1. https://doi.org/10.51291/2377-7478.1691

Cruz, S., Calado, R., Serôdio, J., & Cartaxana, P. (2013). Crawling leaves: photosynthesis in sacoglossan sea slugs. *Journal of Experimental Botany* 64, 3999–4009. https://doi.org/10.1093/jxb/ert197

Cruz, S. & Cartaxana, P. (2022). Kleptoplasty: getting away with stolen chloroplasts. *PLoS Biology* 20, e3001857. https://doi.org/10.1371/journal.pbio.3001857

Cryan, J. F. & Dinan, T. G. (2012). Mind-altering microorganisms: the impact of the gut microbiota on brain and behaviour. *Nature Reviews Neuroscience* 13, 701–712. https://doi.org/10.1038/nrn3346

Damasio, A. (2021). *Feeling and Knowing: Making Minds Conscious*. New York: Pantheon Books.

Damasio, A. (2022). Perspectives on the cognitive unconscious. In: Reber, A. S. & Allen, R. (Eds.), *The Cognitive Unconscious: The First Half-Century*, pp. 366–372. Oxford: Oxford University Press.

Damer, B. (2016). A field trip to the Archaean in search of Darwin's warm little pond. *Life* 6, 21.

da Silva, R. M. F. L. (2014). Syncope: epidemiology, etiology, and prognosis. *Frontiers of Physiology* 5, 471. https://doi.org/10.3389/fphys.2014.00471

Davies, P. (1999). *The Fifth Miracle: The Search for the Origin and Meaning of Life*. New York: Simon and Shuster.

Davy, S. K., Allemand, D., & Weis, V. M. (2012). Cell biology of cnidarian-dinoflagellate symbiosis. *Microbiology and Molecular Biology Reviews* 76, 229–261. https://doi.org/10.1128/MMBR.05014-11

Dawkins, R. (1976/1989). *The Selfish Gene*. New York: Oxford University Press.

Deacon, T. W. (2011). *Incomplete Nature: How Mind Emerged from Matter*. New York: W. W. Norton.

Deamer, D. (2020). Salty seawater or a warm little pond: where did life begin? *Academia Letters* 2. https://www.academia.edu/44977092/Salty_seawater_or_a_warm_little_pond_Where_did_life_begin

de Carpentier, F., Lemaire, S. D., & Danon, A. (2019). When unity is strength: the strategies used by Chlamydomonas to survive environmental stresses. *Cells* 8, 1307. https://doi.org/10.3390/cells8111307

de Carpentier, F., Maes, A., Marchand, C. H., Chung, C., Durand, C., Crozet, P., et al. (2022). How abiotic stress-induced socialization leads to the formation of massive aggregates in *Chlamydomonas*. *Plant Physiology* 190, 1927–1940. https://doi.org/10.1093/plphys/kiac321

De Klerk, G. J. M. (1979). Mechanism and vitalism. A history of the controversy. *Acta Biotheoretica* 28, 1–10.

Delarue, M., Brittingham, G. P., Pfeffer, S., Surovtsev, I. V., Pinglay, S., Kennedy, K. J., et al. (2018). mTORC1 controls phase separation and the biophysical properties of the cytoplasm by tuning crowding. *Cell* 174, 338–349. https://doi.org/10.1016/j.cell.2018.05.042

Delbrück, M. (1949/1999). A physicist looks at biology. *Resonance* 4, 89–102.

De Loof, A. (2015). From Darwin's on the origin of species by means of natural selection to the evolution of life with communication activity as its very essence and driving force (= mega-evolution). *Life: The Excitement of Biology* 3(3).

De Loof, A. (2016). The cell's self-generated 'electrome': the biophysical essence of the immaterial dimension of Life? *Communicative & Integrative Biology* 9, e1197446. https://doi.org/10.1080/19420889.2016.1197446

Demidchik, V. & Shabala, S. (2018). Mechanisms of cytosolic calcium elevation in plants: the role of ion channels, calcium extrusion systems and NADPH oxidase-mediated 'ROS-Ca^{2+} Hub'. *Functional Plant Biology* 45, 9–27. https://doi.org/10.1071/FP16420

Dennett, D. C. (2017). *From Bacteria to Bach and Back: The Evolution of Minds*. New York: Norton.

De Palma, A. & Pareti, G. (2011). Bernstein's long path to membrane theory: radical change and conservation in nineteenth-century German electrophysiology. *Journal of the History of the Neurosciences* 20, 306–337. https://doi.org/10.1080/0964704X.2010.532024

Derbyshire, S. W. G. (2016). Fish lack the brains and the psychology for pain. *Animal Sentience* 3, 18.

Derr, J. B., Tamayo, J., Clark, J. A., Morales, M., Mayther, M. F., Espinoza, E. M., et al. (2020). Multifaceted aspects of charge transfer. *Physical Chemistry Chemical Physics* 22, 21583–21629. https://doi.org/10.1039/d0cp01556c

De Waal, F. (2016). *Are We Smart Enough to Know How Smart Animals Are*? New York: Norton.

Dexter, J. P., Prabakaran, S., & Gunawardena, J. (2019). A complex hierarchy of avoidance behaviors in a single-cell eukaryote. *Current Biology* 29, 4323–4329. https://doi.org/10.1016/j.cub.2019.10.059

Díaz-Garrido, N., Badia, J., & Baldomà, L. (2021). Microbiota-derived extracellular vesicles in interkingdom communication in the gut. *Journal of Extracellular Vesicles* 10, e12161. https://doi.org/10.1002/jev2.12161

Diaz-Munoz, S. L., Sanjuan, R., & West, S. (2017). Sociovirology: conflict, cooperation, and communication among viruses. *Cell Host & Microbe* 22, 437–441. https://doi.org/10.1016/j.chom.2017.09.012

Di Giulio, M. (2021). The late appearance of DNA, the nature of the LUCA and ancestors of the domains of life. *Biosystems* 202, 104330. https://doi.org/10.1016/j.biosystems.2020.104330

Diering, G. H., Mills, F., Bamji, S. X., & Numata, M. (2011). Regulation of dendritic spine growth through activity-dependent recruitment of the brain-enriched Na^+/H^+ exchanger NHE5. *Molecular Biology of the Cell* 22, 2246–2257. https://doi.org/10.1091/mbc.E11-01-0066

Diering, G. H. & Numata, M. (2014). Endosomal pH in neuronal signaling and synaptic transmission: role of Na+/H+ exchanger NHE5. *Frontiers in Physiology* 4, 412. https://doi.org/10.3389/fphys.2013.00412

Dillard, M. M. (1930). Ethylene—the new general anesthetic. *Journal of National Medical Association* 1, 10–11. PMCID: PMC2624845

Dirk, J. A. & Verkman, A. S. (2008). Crowding effects on diffusion in solutions and cells. *Annual Reviews of Biochemistry* 37, 247–263. https://doi.org/10.1146/annurev.biophys.37.032807.125824

Dodig-Crnkovic, G. (2022). Cognition as morphological/morphogenetic embodied computation in vivo. *Entropy* 24, 1576. https://doi.org/10.3390/e24111576

Donau, C., Späth, F., Sosson, M., Kriebisch, B. A. K., Schnitter, F., Tena-Solsona, M., et al. (2020). Active coacervate droplets as a model for membraneless organelles and protocells. *Nature Communications* 11, 52167. https://doi.org/10.1038/s41467-020-18815-9

Dong, T., Li, J., Liao, Y., Chen, B. J. W., & Xu, X. (2017). Root-mediated sex recognition in a dioecious tree. *Scientific Reports* 7, 801. https://doi.org/10.1038/s41598-017-00894-2

Draguhn, A., Mallatt, J. M., & Robinson, D. G. (2021). Anesthetics and plants: no pain, no brain, and therefore no consciousness. *Protoplasma* 258, 239–248. https://doi.org/10.1007/s00709-020-01550-9

Dretske, F. I. (1981). *Knowledge and the Flow of Information*. Cambridge, MA: MIT Press.

Drongitis, D., Aniello, F., Fucci, L., & Donizetti, A. (2019). Roles of transposable elements in the different layers of gene expression regulation. *International Journal of Molecular Sciences* 20, 5755. https://doi.org/10.3390/ijms20225755

Dupré, J. & O'Malley, M. A. (2009). Varieties of living things: life at the intersection of lineage and metabolism. *Philosophicla & Theoretical Biology* 1, e003. https://doi.org/10.1007/978-94-007-2445-7_13

Dussutour, A. (2021). Learning in single cell organisms. *Biochemical Biophysics Research Communications* 564, 92–102. https://doi.org/10.1016/j.bbrc.2021.02.018

Eckenhoff, R. G. (2008). Why can all biology be anesthesized? *Anesthesia & Anelgesia* 107, 859–861. https://doi.org/10.1213/ane.0b013e31817ee7ee

Edgar, R. S., Green, E. W., Zhao, Y., van Ooijen, G., Olmedo, M., Qin, X., et al. (2012). Peroxiredoxins are conserved markers of circadian rhythms. *Nature* 485, 459–464. https://doi.org/10.1038/nature11088

Edidin, M. (2003). The state of lipid rafts: from model membranes to cells. *Annual Reviews of Biophysics and Biomolecular Structure* 32, 257–283. https://doi.org/10.1146/annurev.biophys.32.110601.142439

Edwards, J. C. W. (2005). Is consciousness only a property of individual cells? *Journal of Consciousness Studies* 12, 60–76.

Eger, E. I., 2nd, Raines, D. E., Shafer, S. L., Hemmings. H. C., Jr., & Sonner, J. M. (2008). Is a new paradigm needed to explain how inhaled anesthetics produce immobility. *Anesthesia & Analgesia* 107, 832–848. https://doi.org/10.1213/ane.0b013e318182aedb

Eger, II E. I., Saidman, L. J., & Wethorpe, R. N. (2014b). 1844–1846: The discovery and demonstration of anesthesia. In: Eger, II E. I., Saidman, L. J., & Wethorpe, R. N. (Eds.), *The Wondrous Story of Anesthesia*, pp. 17–26. Springer New York Heidelberg Dordrecht London. https://doi.org/10.1007/978-1-4614-8441-7_1

Eisenstein, E. M. & Eisenstein, D. (2006). A behavioral homeostasis theory of habituation and sensitization: II. Further developments and predictions. *Reviews in the Neurosciences* 17, 533–558. https://doi.org/10.1515/revneuro.2006.17.5.533

Ellis, R. J. (2001). Macromolecular crowding: an important but neglected aspect of the intracellular environment. *Current Opinions of Structural Biology* 11, 114–119. https://doi.org/10.1016/S0959-440X(00)00172

Elsaesser, T. (2009). Ultrafast memory loss and relaxation processes in hydrogen-bonded systems. *Biological Chemistry* 390, 1125–1132. https://doi.org/10.1515/BC.2009.126

Elsasser, W. M. (1981). Principles of a new biological theory: a summary. *Journal of Theoretical Biology* 89, 131–150. https://doi.org/10.1016/0022-5193(81)90182-x

Elsasser, W. M. (1987). *Reflections on a Theory of Organisms: Holism in Biology*. Baltimore, MD: Johns Hopkins University Press.

Emenecker, R. J., Holehouse, A. S., & Strader, L. C. (2021). Biological phase separation and biomolecular condensates in plants. *Annual Reviews of Plant Biology* 72, 17–46. https://doi.org/10.1146/annurev-arplant-081720-015238

Emmeche, C. (2004). Organicism and qualitative aspects of self-organization. *Revue Internationale de Philosophie* 228, 205–217. https://doi.org/10.3917/rip.228.0205

Epstein, R. (2016). The empty brain. *Aeon* 18 May, 2016. https://aeon.co/essays/your-brain-does-not-process-information-and-it-is-not-a-computer

Ettema, T. J., Lindås, A. C., & Bernander, R. (2011). An actin-based cytoskeleton in archaea. *Molecular Microbiology* 80, 1052–1061. https://doi.org/10.1111/j.1365-2958.2011.07635.x

Evans, M. J., Choi, W. G., Gilroy, S., & Morris, R. J. (2016). A ROS-assisted calcium wave dependent on the AtRBOHD NADPH oxidase and TPC1 cation channel propagates the systemic response to salt stress. *Plant Physiology* 171, 1771–1784. https://doi.org/10.1104/pp.16.00215

Ewart, A. J. (1903). *On the Physics and Physiology of Protoplasmic Streaming in Plants*. Oxford: Clarendon Press.

Fabricant, A., Iwata, G. Z., Scherzer, S., Bougas, L., Rolfs, K., Jodko-Władzińska, A., et al. (2021). Action potentials induce biomagnetic fields in carnivorous Venus flytrap plants. *Scientific Reports* 11, 1438. https://doi.org/10.1038/s41598-021-81114-w

Fadloun, A., Le Gras, S., Jost, B., Ziegler-Birling, C., Takahashi, H., Gorab, E., et al. (2013). Chromatin signatures and retrotransposon profiling in mouse embryos reveal regulation of LINE-1 by RNA. *Nature Structural & Molecular Biology* 20, 332–338. https://doi.org/10.1038/nsmb.2495

Fan, S. H. Y., Numata, Y., & Numata, M. (2016). Endosomal Na+/H+ exchanger NHE5 influences MET recycling and cell migration. *Molecular Biology of the Cell* 27, 702–715. https://doi.org/10.1091/mbc.E15-04-0257

Fang, Y., Wang, Z., Liu, X., & Tyler, B. M. (2022). Biogenesis and biological functions of extracellular vesicles in cellular and organismal communication with microbes. *Frontiers of Microbiology* 13, 817844. https://doi.org/10.3389/fmicb.2022.817844

Farnsworth, K. D., Nelson, J., & Gershenson, C. (2013). Living is information processing: from molecules to global systems. *Acta Biotheoretica* 61, 203–222. https://doi.org/10.1007/s10441-013-9179-3

Fatema, U., Ali, M. F., Hu, Z., Clark, A. J., & Kawashima, T. (2019). Gamete nuclear migration in animals and plants. *Frontiers in Plant Science* 10, 517. https://doi.org/10.3389/fpls.2019.00517

Fayer, M. D. (2011). Water in a crowd. *Physiology (Bethesda)* 26, 381–392. https://doi.org/10.1152/physiol.00021.2011

Fayer, M. D. (2012). Dynamics of water interacting with interfaces, molecules, and ions. *Accounts of Chemical Research* 45, 3–14. https://doi.org/10.1021/ar2000088

Fedorenko, O. A. & Marchenko, S. M. (2014). Ion channels of the nuclear membrane of hippocampal neurons. *Hippocampus* 24, 869–876. https://doi.org/10.1002/hipo.22276

Fedoroff, N. V. (2012). Transposable elements, epigenetics, and genome evolution. *Science* 338, 758–767. https://www.science.org/doi/10.1126/science.338.6108.758

Feinberg, T. E. & Mallatt, J. (2013). *The Ancient Origins of Consciousness: How the Brain Created Experience*. Cambridge, MA: MIT Press.

Feng, Z., Chen, X., Wu, X., & Zhang, M. (2019). Formation of biological condensates via phase separation: characteristics, analytical methods, and physiological implications. *Journal of Biological Chemistry* 294, 14823–14835. https://doi.org/10.1074/jbc.REV119.007895

Ferdousi, M. & Finn, D. P. (2018). Stress-induced modulation of pain: role of the endogenous opioid system. *Progress in Brain Research* 239, 121–177. https://doi.org/10.1016/bs.pbr.2018.07.002

Fichman, Y. & Mitler, R. (2021). Integration of electric, calcium, reactive oxygen species and hydraulic signals during rapid systemic signaling in plants. *Plant Journal* 107, 7–20. https://doi.org/10.1111/tpj.15360

Fichtl, B., Shrivastava, S., & Schneider, M. F. (2016). Protons at the speed of sound: predicting specific biological signaling from physics. *Scientific Reports* 6, 22874. https://doi.org/10.1038/srep2287

Fields, C., Glazebrook, J. F., & Levin, M. (2021). Minimal physicalism as a scale-free substrate for cognition and consciousness. *Neuroscience of Consciousness* 2021, niab013. https://doi.org/10.1093/nc/niab013

Fields, C., Glazebrook, J. F., & Levin, M. (2022). Neurons as hierarchies of quantum reference frames. *Biosystems* 219, 104714. https://doi.org/10.1016/j.biosystems.2022.104714

Fields, C. & Levin, M. (2018). Multiscale memory and bioelectric error correction in the cytoplasm-cytoskeleton-membrane system. *Wiley Interdisciplinary Reviews* 10, e1410. https://doi.org/10.1002/wsbm.1410

Fields, C. & Levin, M. (2021). Metabolic limits on classical information processing by biological cells. *BioSystems* 209, 104513. https://doi.org/10.1016/j.biosystems.2021.104513

Fillafer, C., Paeger, A., & Schneider, M. F. (2021). The living state: how cellular excitability is controlled by the thermodynamic state of the membrane. *Progress in Biophysics and Molecular Biology* 162, 57–68. https://doi.org/10.1016/j.pbiomolbio.2020.10.003

Fillingame, R. H. (2000). Getting to the bottom of the F1-ATPase. *Nature Structural Biology* 7, 1002–1004. https://doi.org/10.1038/80902

Finlay, B. J. (2004). Protist taxonomy: an ecological perspective. *Proceedings of the Royal Society B—Biological Sciences* 359, 599–610. https://doi.org/10.1098/rstb.2003.1450

Fitch, W. T. (2008). Nano-intentionality: a defense of intrinsic intentionality. *Biology & Philosophy* 23, 157–177. https://doi.org/10.1007/s10539-007-9079-5

Floridi, L. (2003). From data to semantic information. *Entropy* 5, 125–145.

Foffi, G., Pastore, A., Piazza, F., & Temussi, P. A. (2013). Macromolecular crowding: chemistry and physics meet biology (Ascona, Switzerland, 10–14 June 2012). *Physical Biology* 10, 040301. https://doi.org/10.1088/1478-3975/10/4/040301

Ford, B. J. (2004). Are Cells Ingenious? *Microscope* 52(3/4), 135–144.

Ford, B. J. (2009). On intelligence in cells: the case for whole cell biology. *Interdisciplinary Science Reviews* 34, 350–365. https://doi.org/10.1179/030801809X12529269201282

Ford, B. J. (2017). Cellular intelligence: microphenomenology and the realities of being. *Progress in Biophysics and Molecular Biology* 131, 273–287. https://doi.org/10.1016/j.pbiomolbio.2017.08.012

Forgac, M. (2007). Vacuolar ATPases: rotary proton pumps in physiology and pathophysiology. *Nature Reviews Molecular and Cell Biology* 8, 917–929. https://doi.org/10.1038/nrm2272

Forshaw, S. D. (2016). The third state: toward a quantum information theory of consciousness. *NeuroQuantology* 14, 49–61. https://doi.org/10.14704/nq.2016.14.1.899

Fournier, G., Moore, K. R., Rangel, L. T., Payette, J. G., Momper, L., & Bosak, T. (2021). The Archean origin of oxygenic photosynthesis and extant cyanobacterial lineages. *Proceedings of the Royal Society B—Biological Sciences* 288, 2021675. https://doi.org/10.1098/rspb.2021.0675

Francis, J. & Dingley, J. (2015). Electroanaesthesia—from torpedo fish to TENS. *Anesthesia* 70, 93–103. https://doi.org/10.1111/anae.12887

Frank, A., Grinspsoon, D., & Walker, S. (2022). Intelligence as a planetary scale process. *International Journal of Astrobiology* 21, 47–61.

Frank, J., Happeck, R., Meier, B., Hoang, M. T. T., Stribny, J., Hause, G., et al. (2018). Chloroplast-localized BICAT proteins shape stromal calcium signals and are required for efficient photosynthesis. *New Phytologist* 221, 866–880. https://doi.org/10.1111/nph.15407

Freddolino, P. L., Yang, J., Momen-Roknabadi, A., and Tavazoie, S. (2018). Stochastic tuning gene expression enables cellular adaptation in the absence of pre-existing regulatory circuitry. *eLife* 7, e31867. https://doi.org/10.7554/eLife.31867

Fukasawa, Y., Savoury, M., & Boddy, L. (2020). Ecological memory and relocation decisions in fungal mycelial networks: responses to quantity and location of new resources. *ISME Journal* 14, 380–388. https://doi.org/10.1038/s41396-019-0536-3

Gabor, D. 1946. Theory of communication. Part 1: the analysis of information. *Journal of the Institution of Electrical Engineers Part III: Radio and Communication Engineering* 93, 429–441.

Gagliano, M., Mancuso, S., & Robert, D. (2012). Towards understanding plant bioacoustics. *Trends in Plant Science* 17, 323–325. https://doi.org/10.1016/j.tplants.2012.03.002

Gagliano, M., Renton, M., Depczynski, M., & Mancuso, S. (2014). Experience teaches plants to learn faster and forget slower in environments where it matters. *Oecologia* 175, 63–72. https://doi.org/10.1007/s00442-013-2873-7

Galassi, V. V. & Wilke, N. (2021). On the coupling between mechanical properties and electrostatics in biological membranes. *Membranes (Basel)* 11, 478. https://doi.org/10.3390/membranes11070478

Galera-Laporta, L., Comerci, C. J., Garcia-Ojalvo, J., & Süel, G. M. (2021). IonoBiology: the functional dynamics of the intracellular metallome, with lessons from bacteria. *Cell Systems* 12, 497–508. https://doi.org/10.1016/j.cels.2021.04.011

Galperin, M. Y. (2005). Life is not defined just in base pairs. *Environmental Microbiology* 7, 149–152. https://doi.org/10.1111/j.1462-2920.2005.00774.x

Gao, A. Y., Lourdin-De Filippis, E., Orlowski, J., & McKinne, R. A. (2022). Roles of endomembrane alkali cation/proton exchangers in synaptic function and neurodevelopmental disorders. *Frontiers in Physiology* 13, 892196. https://doi.org/10.3389/fphys.2022.892196

Garcia-Cabau, C. & Salvatella, X. (2021). Regulation of biomolecular condensate dynamics by signaling. *Current Opinion in Cell Biology* 69, 111–119. https://doi.org/10.1016/j.ceb.2021.01.002

García-Sierra, F. & Frixione, E. (1993). Lidocaine, a local anesthetic, reversibly inhibits cytoplasmic streaming in *Valisneria* mesophyll cells. *Protoplasma* 175, 153–160. https://doi.org/10.1007/BF01385014

Gaskett, A. C., Winnick, C. G., & Herberstein, M. E. (2008). Orchid sexual deceit provokes ejaculation. *American Naturalist* 171, E206–E212. https://doi.org/10.1086/587532

Gavelis, G. S., Hayakawa, S., White III, R. A., Gojobori, T., Suttle, C. A., Keeling, P. J., et al. (2015). Eye-like ocelloids are built from different endosymbiotically acquired components. *Nature* 523, 204–207. https://doi.org/10.1038/nature14593

George, B. M. & Pandit, J. J. (2021). General anaesthetics as 'awakening agents'? Re-appraising the evidence for suggested 'pressure reversal' of anaesthesia. *Clinical and Experimental Pharmacology and Physiology* 48, 1454–1468. https://doi.org/10.1111/1440-1681.13554

Ghosh, S., Banerjee-Ghosh, K., Levy, D., Scheerer, D., Riven, I., Shin, J., et al. (2022). Control of protein activity by photoinduced spin polarized charge reorganization. *Proceedings of the National Academy of Sciences USA* 119, e2204735119. https://doi.org/10.1073/pnas.2204735119

Gianoli, E. (2017). Eyes in the chameleon vine? *Trends in Plant Science* 22, 4–5. https://doi.org/10.1016/j.tplants.2016.11.001

Gianoli, E. & Carrasco-Urra, F. (2014). Leaf mimicry in a climbing plant protects against herbivory. *Current Biology* 24, 984–987. https://doi.org/10.1016/j.cub.2014.03.010

Gianoli, E., González-Teuber, M., Vilo, C., Guevara-Araya, M. J., & Escobedo, V. M. (2021). Endophytic bacterial communities are associated with leaf mimicry in the vine *Boquila trifoliolata*. *Scientific Reports* 11, 22673. https://doi.org/10.1038/s41598-021-02229-8

Gilbert, C. & Cordaux, R. (2017). Viruses as vectors of horizontal transfer of genetic material in eukaryotes. *Current Opinions in Virology* 25, 16–22. https://doi.org/10.1016/j.coviro.2017.06.005

Gilbert, S. F., McDonald, E., Boyle, N., Buttino, N., Gyi, L., Mai, M., et al. (2010). Symbiosis as a source of selectable epigenetic variation: taking the heat for the big guy. *Philosophical Transactions of the Royal Society B: Biological Sciences* 365, 671–678. https://doi.org/10.1098/rstb.2009.0245

Gilbert, S. F. & Sarkar, S. (2000). Embracing complexity: organicism for the 21st century. *Developmental Dynamics* 219, 1–9.

Gillen, A. L. & Sherwin, F. (2008). Louis Pasteur's views on creation, evolution, and the genesis of germs. *Answers Research Journal* 1, 43–52.

Gilroy, S., Białasek, M., Suzuki, N., Górecka, M., Devireddy, A. R., Karpiński, S., et al. (2016). ROS, calcium, and electric signals: key mediators of rapid systemic signaling in plants. *Plant Physiology* 171, 1606–1615. https://doi.org/10.1104/pp.16.00434

Gilroy, S., Suzuki, N., Miller, G., Choi, W. G., Toyota, M., Devireddy, A. R., et al. (2014). A tidal wave of signals: calcium and ROS at the forefront of rapid system. *Trends in Plant Sciences* 19, 623–630. https://doi.org/10.1016/j.tplants.2014.06.013

Gilroy, S. & Trewavas, A. (2022). Agency, teleonomy and signal transduction in plant systems. *Biological Journal of the Linnean Society* blac021. https://doi.org/10.1093/biolinnean/blac021

Ginsburg, S. & Jablonka, E. (2019). *The Evolution of the Sensitive Soul: Learning and the Origins of Consciousness*. Cambridge, MA: MIT Press.

Giovannetti, M., Avio, L., Fortuna, P., Pellegrino, E., Sbrana, C., & Strani, P. (2006). At the root of the wood wide web—self recognition and nonself incompatibility in mycorrhizal networks. *Plant Signaling & Behavior* 1, 1–5. https://doi.org/10.4161/psb.1.1.2277

Glasgow, R. (2018). *Minimal Selfhood and the Origins of Consciousness*. Würzburg: Würzburg University Press.

Godfrey-Smith, P. (2016). *Other Minds: The Octopus, the Sea, and the Deep Origins of Consciousness*. New York: Farrar, Straus & Giroux.

Goldenfeld, N., & Woese, C. (2007). Biology's next revolution. *Nature* 445, 369.

Goldsworthy, A. (1983). The evolution of plant action potentials. *Journal of Theoretical Biology* 103, 645–648. https://doi.org/10.1023/A:1005020826000

Gómez-Márquez, J. (2020). What are the principles that govern life? *Communicative & Integrative Biology* 13, 97–107. https://doi.org/10.1080/19420889.2020.1803591

González-Teuber, M. & Heil, M. (2009). Nectar chemistry is tailored for both attraction of mutualists and protection from exploiters. *Plant Signaling & Behavior* 4, 809–813. https://doi.org/10.4161/psb.4.9.9393

Gooday, A. J., Durgen, J. M., & Smith, C. R. (2020). Giant, highly diverse protists in the abyssal Pacific: vulnerability to impacts from seabed mining and potential for recovery. *Communicative & Integrative Biology* 13, 189–197. https://doi.org/10.1080/19420889.2020.1843818

Gooday, A. J., Schoenle, A., Dolan, J. R., & Arndt, H. (2021). Protist diversity and function in the dark ocean—challenging the paradigms of deep-sea ecology with special emphasis on foraminiferans and naked protists. *European Journal of Protistology* 75, 125721. https://doi.org/10.1016/j.ejop.2020.125721

Gopal, S. K., Greening, D. W., Rai, A., Chen, M., Xu, R., Shafiq, A., et al. (2017). Extracellular vesicles: their role in cancer biology and epithelial-mesenchymal transition. *Biochemical Journal* 474, 21–45. https://doi.org/10.1042/BCJ20160006

Gorham, G. (1994). Mind-body dualism and the Harvey-Descartes controversy. *Journal of the History of Ideas* 55, 211–234. https://doi.org/10.2307/2709897

Gould, S. J. (2002). *The Structure of Evolutionary Theory*. Cambridge, MA: Belknap Press.

Gözen, I., Köksal, E. S., Põldsalu, I., Xue, L., Spustova, K., Pedrueza-Villalmanzo, E., et al. (2022). Protocells: milestones and recent advances. *Small* 18, e2106624. https://doi.org/10.1002/smll.202106624

Granak, S., Hoschl, C., & Ovsepian, S. V. (2021). Dendritic spine remodeling and plasticity under general anesthesia. *Brain Structure and Function* 226, 2001–2017. https://doi.org/10.1007/s00429-021-02308-6

Grasso, D. A., Pandolfi, C., Bazihizina, N., Nocentini, D., Nepi, M., & Mancuso, S. (2015). Extrafloral-nectar-based partner manipulation in plant-ant relationships. *AoB Plants* 7, plv002. https://doi.org/10.1093/aobpla/plv002

Gray, M. W. (2017). Lynn Margulis and the endosymbiont hypothesis: 50 years later. *Molecular Biology of the Cell* 28, 1285–1287. https://doi.org/10.1091/mbc.E16-07-0509

Grémiaux, A., Yokawa, K., Mancuso, S., & Baluška, F. (2014). Plant anesthesia supports similarities between animals and plants: Claude Bernard's forgotten studies. *Plant Signaling & Behavior* 9, e27886. https://doi.org/10.4161/psb.27886

Griffin, D. R. (1992). *Animal Minds: Beyond Cognition to Consciousness*. Chicago, IL: University of Chicago Press.

Grimes, P. J., Galanti, A., & Gobbo, P. (2021). Bioinspired networks of communicating synthetic protocells. *Frontiers in Molecular Biosciences* 8, 804717. https://doi.org/10.3389/fmolb.2021.804717

Grüber, G., Manimekalai, M. S. S., Mayer, F., & Müller, V. (2014). ATP synthases from archaea: the beauty of a molecular motor. *Biochimica et Biophysica Acta (BBA)—Bioenergetics* 6, 940–952. https://doi.org/10.1016/j.bbabio.2014.03.004

Gruntman, M. & Novoplansky, A. (2004). Physiologically mediated self/non-self discrimination. *Proceedings of the National Academy of Sciences USA* 101, 3863–3867. https://doi.org/10.1073/pnas.0306604101

Hahn, A., Vonck, J., Mills, D. J., Meier, T., & Kühlbrandt, W. (2018). Structure, mechanism, and regulation of the chloroplast ATP synthase. *Science* 360, eaat4318. https://doi.org/10.1126/science.aat4318

Hales, C. G. & Ericson, M. (2022). Electromagnetism's bridge across the explanatory gap: how a neuroscience/physics collaboration delivers explanation into all theories of consciousness. *Frontiers in Human Neuroscience* 16, 836046. https://doi.org/10.3389/fnhum.2022.836046

Hameroff, S. R. (2006). The entwined mysteries of anesthesia and consciousness. *Anesthesiology* 105, 400–412. https://doi.org/10.1097/00000542-200608000-00024

Hameroff, S. R. (2021). 'Orch OR' is the most complete, and most easily falsifiable theory of consciousness. *Cognitive Neuroscience* 12, 74–76. https://doi.org/10.1080/17588928.2020.1839037

Hameroff, S. & Penrose, R. (2014). Consciousness in the universe: a review of the 'Orch OR' theory. *Physics of Life Reviews* 11, 39–78. https://doi.org/10.1016/j.plrev.2013.08.002

Händeler, K., Grzymbowski, Y. P., Krug, P. J., & Wägele, H. (2009). Functional chloroplasts in metazoan cells—a unique strategy in animal life. *Frontiers in Zoology* 6, 28–45. https://doi.org/10.1186/1742-9994-6-28

Hantak, M. P., Einstein, J., Kearns, R. B., & Shepherd, J. D. (2021). Intercellular communication in the nervous system goes viral. *Trends in Neurosciences* 44, 248–259. https://doi.org/10.1016/j.tins.2020.12.003

Hao, Z., Liu, J., Wu, B., Yu, M., & Wegner, L. H. (2021). Strong emergence in biological systems: is it open to mathematical reasoning? *Acta Biotheoretica* 69, 841–856. https://doi.org/10.1007/s10441-021-09423-1

Hardeland, R. (2022). Redox biology of melatonin: discriminating between circadian and non-circadian functions. *Antioxidand and Redox Signaling* 37, 704–725. https://doi.org/10.1089/ars.2021.0275

Hardman, J. G. (2017). *Oxford Textbook of Anaesthesia*. Oxford: Oxford University Press.

Harel, A., Bromberg, Y., Falkowski, P. G., & Bhattacharya, D. (2018). Evolutionary history of redox metal-binding domains across the tree of life. *Proceedings of the National Academy of Sciences USA* 111, 7044–7047. https://doi.org/10.1073/pnas.1403676111

Harmon, S. H. E. (2009). Of plants and people. Why do we care about dignity? *EMBO Reports* 10, 946–948. https://doi.org/10.1038/embor.2009.191

Harnad, S. (2001). Minds, machines, and Searle 2: what's wrong and right about Searle's Chinese Room argument. In: Bishop, M. & Preston, J. (Eds.), *Views into the Chinese Room: Essays on Searle's Chinese Room Argument*, Chapter 14. Oxford: Oxford University Press.

Harold, F. M. (2022). *On Life: Cells, Genes, and the Evolution of Complexity*. New York: Oxford University Press. https://doi.org/10.1093/oso/9780197604540.001.0001

Hatano, T., Palani, S., Papatziamou, D., Salzer, R., Souza, D. P., Tamarit, D., et al. (2022). Asgard archaea shed light on the evolutionary origins of the eukaryotic ubiquitin-ESCRT machinery. *Nature Communications* 13, 3398. https://doi.org/10.1038/s41467-022-30656-2

Hauser, L. (1997). Searle's Chinese box: debunking the Chinese Room argument. *Minds and Machines*, 7, 199–226.

He, X., Chadwick, G. L., Kempes, C. P., Orphan, V. J., & Meile, C. (2021). Controls on inter-species electron transport and size limitation of anaerobically methane-oxidizing microbial consortia. *mBio* 12, e03620–20. https://doi.org/10.1128/mBio.03620-20

He, Y. & Li, Z. (2018). Epigenetic environmental memories in plants: establishment, maintenance and reprogramming. *Trends in Genetics* 34, 856–866. https://doi.org/10.1016/j.tig.2018.07.006

Hedges, D. J. & Belancio, V. P. (2011). Restless genomes: humans as a model organism for understanding host-retrotransposable element dynamics. *Advances in Genetics* 73, 219–262. https://doi.org/10.1016/B978-0-12-380860-8.00006-9

Hedrich, R. (2012). Ion channels in plants. *Physiology Reviews* 92, 1777–1811. https://doi.org/10.1152/physrev.00038.2011

Hedrich, R., Salvador-Recatalà, V., & Dreyer, I. (2016). Electrical wiring and long-distance plant communication. *Trends in Plant Science* 21, 376–387. https://doi.org/10.1016/j.tplants.2016.01.016

Helman, P. (1986). The principle of optimality in the design of efficient algorithms. *Journal of Mathematical Analysis and Applications* 119, 97–127.

Hemmings, H. C., Jr., Yan, W., Westphalen, R. I., & Ryan, T. A. (2005). The general anesthetic isoflurane depresses synaptic vesicle exocytosis. *Molecular Pharmacology* 67, 1591–1599. https://doi.org/10.1124/mol.104.003210

Hempel de Ibarra, N. & Rands, S. (2021). Pollination: influencing bee behaviour with caffeine. *Current Biology* 31, R1090–R1092. https://doi.org/10.1016/j.cub.2021.07.073

Hendricks, S. B. & Taylorson, R. B. (1979). Overcoming dormancy in seeds with ethanol and other anesthetics. *Planta* 145, 507–510. https://doi.org/10.1007/BF00380106

Hendricks, S. B. & Taylorson, R. B. (1980). Reversal by pressure of seed germination promoted by anesthetics. *Planta* 149, 108–111. https://doi.org/10.1007/BF00380869

Herron, M. D. (2016). Origins of multicellular complexity: *Volvox* and the volvocine algae. *Molecular Ecology* 25, 1213–1223. https://doi.org/10.1111/mec.13551

Herron, M. D., Borin, J. M., Boswell, J. C., Walker, J., Chen, I. C. K., Knox, C. A., et al. (2019). De novo origins of multicellularity in response to predation. *Scientific Reports* 9, 2328. https://doi.org/10.1038/s41598-019-39558-8

Herron, M. D., Hackett, J. D., Aylward, F. O., & Michod, R. E. (2009). Triassic origin and early radiation of multicellular volvocine algae. *Proceedings of the National Academy of Sciences USA* 106, 3254–3258. https://doi.org/10.1073/pnas.0811205106

Hijaz, F. & Killiny, N. (2019). Exogenous GABA is quickly metabolized to succinic acid and fed into the plant TCA cycle. *Plant Signaling & Behavior* 14, e1573096. https://doi.org/10.1080/15592324.2019.1573096

Hille, B. (2022). Ionic channels in nerve membranes, 50 years on. *Progress in Biophysics and Molecular Biology* 169-170, 12–20. https://doi.org/10.1016/j.pbiomolbio.2021.11.003

Hirsch, C. D. & Springer, N. M. (2017). Transposable element influences on gene expression in plants. *Biochimica et Biophysica Acta (BBA)-Gene Regulatory Mechanisms* 1860, 157–165.

Ho, M.-W.-W. (1994). *The Rainbow and the Worm: The Physics of Organisms.* Singapore: World Scientific.

Hoffmann, P. M. (2012). *Life's Ratchet: How Molecular Machines Extract Order from Chaos.* New York: Basic Books.

Hoffmann, P. M. (2016). How molecular motors extract order from chaos. *Reports on Progress in Physics* 79, 032601. https://doi.org/10.1088/0034-4885/79/3/032601

Hollants, J., Decleyre, H., Leliart, F., Dr Clerk, O., & Willemn, A. (2011). Life without a cell membrane: challenging the specificity of bacterial endophytes within *Bryopsis* (Bryopsidales, Chlorophyta). *BMC Microbiology* 11, 255. https://doi.org/10.1186/1471-2180-11-255

Honig, B. H., Hubbell, W. L., & Flewelling, R. F. (1986). Electrostatic interactions in membranes and proteins. *Annual Review of Biophysics and Biophysical Chemistry* 15, 163–193. https://doi.org/10.1146/annurev.bb.15.060186.001115

Hooper, L. V., Littman, D. R., & Macpherson, A. J. (2012). Interactions between the microbiota and the immune system. *Science* 336, 1268–1273. https://doi.org/10.1126/science.1223490

Howard, S. R., Greentree, J., Avarguès-Weber, A., Garcia, J. E., Greentree, A. D., & Dyer, A. G. (2022). Numerosity categorization by parity in an insect and simple neural network. *Frontiers in Ecological Evolution* 10, 805385. https://doi.org/10.3389/fevo.2022.805385

Huang, Z., Kelly, S., Matsuo, R., Li, L. F., Li, Y., Olsen, K. M., et al. (2018). The role of standing variation in the evolution of weedines traits in South Asian weedy rice (*Oryza* spp.). *G3 (Bethesda)* 8, 3679–3690. https://doi.org/10.1534/g3.118.200605

Hua-Van, A., Le Rouzic, A., Boutin, T. S., Filée, J., & Capy, P. (2011). The struggle for life of the genome's selfish architects. *Biology Direct* 6, 1–29. https://doi.org/10.1186/1745-6150-6-19

Humphries, C. (2013). Life's beginnings: studying how life bloomed on Earth- and might emerge elsewhere. *Harvard Magazine* September–October 29–34. https://www.harvardmagazine.com/sites/default/files/pdf/2013/09-pdfs/0913-29.pdf

Humphreys, R. K. & Ruxton, G. D. (2018). A review of thanatosis (death feigning) as an anti-predator behaviour. *Behavioral Ecology and Sociobiology* 72, 22. https://doi.org/10.1007/s00265-017-2436-8

Humphries, J., Xiong, L., Liu, J., Prindle, A., Yuan, F., Arjes, H. A., et al. (2017). Species-independent attraction to biofilms through electrical signaling. *Cell* 168, 200–209. https://doi.org/10.1016/j.cell.2016.12.014

Hunley, C. & Marucho, M. (2022). Electrical propagation of condensed and diffuse ions along actin filaments. *Journal of Computational Neuroscience* 50, 91–107. https://doi.org/10.1007/s10827-021-00795-4

Hunt, H. T. (1995). *On the Nature of Consciousness: Cognitive, Phenomenological, and Transpersonal Perspectives*. New Haven, CT: Yale University Press.

Hunt, T., Ericson, M., & Schooler, J. W. (2022). Where's my consciousness-ometer? How to test for the presence and complexity of consciousness. *Perspectives on Psychological Science* 17, 1150–1165. https://doi.org/10.1177/17456916211029942

Hunt, T. & Schooler, J. W. (2019). The easy part of the Hard Problem: a resonance theory of consciousness. *Frontiers in Human Neuroscience* 12, 378. https://doi.org/10.3389/fnhum.2019.00378

Hunter, P. (2021). The emergence of consciousness. *EMBO Reports* 22, e53199. https://doi.org/10.15252/embr.202153199

Hyman, A. A., Weber, C. A., & Jülicher, F. (2014). Liquid-liquid phase separation in biology. *Annual Review of Cell and Developmental Biology* 30, 39–58. https://doi.org/10.1146/annurev-cellbio-100913-013325

Ibuki, T., Imada, K., Minamino, T., Kato, T., Miyata, T., & Namba, K. (2011). Common architecture of the flagellar type III protein export apparatus and F- and V-type ATPase. *Nature Structural & Molecular Biology* 18, 277–282. https://doi.org/10.1038/nsmb.1977

Igamberdiev, A. U. & Shklovskiy-Kordi, N. E. (2017). The quantum basis of spatiotemporality in perception and consciousness. *Progress in Biophysics and Molecular Biology* 130, 15–25.

Ignatov, I. & Mosin, O. V. (2014). Coronal gas discharge effect in modeling of non-equilibrium conditions with gas electric discharge simulating primary atmosphere and hydrosphere for origin of life and living matter. *Journal of Medicine, Physiology and Biophysics* 5, 47–70.

Ikuta, K. S., Swetschinski, L. R., Aguilar, G. R., Sharara, F., Mestrovic, T., Gray, A. P., et al. (2022). Global mortality associated with 33 bacterial pathogens in 2019: a systematic analysis for the Global Burden of Disease Study 2019. *Lancet* 400, 2221–2248. https://doi.org/10.1016/S0140-6736(22)02185-7

Ille, A. M., Lamont, H., & Mathews, M. B. (2022). The central dogma revisited: insights from protein synthesis, CRISPR, and beyond. *Wiley Interdisciplinary Reviews: RNA* 13, e1718.

Imada, K., Minamino, T., Tahara, A., & Namba, K. (2007). Structural similarity between the flagellar type III ATPase FliI and F1-ATPase subunits. *Proceedings of the National Academy of Sciences USA* 104, 485–490. https://doi.org/10.1073/pnas.0608090104

Inaba, K. (2011). Sperm flagella: comparative and phylogenetic perspectives of protein components. *Molecular Human Reproduction* 17, 524–538. https://doi.org/10.1093/molehr/gar034

Irwin, N. A. T., Pittis, A. A., Richards, T. A., & Keeling, P. J. (2022). Systematic evaluation of horizontal gene transfer between eukaryotes and viruses. *Nature Microbiology* 7, 327–336. https://doi.org/10.1038/s41564-021-01026-3

Ishikawa, M., Yuyama, I. Shimizu, H., Nozawa, M., Ikeo, K., & Gojobori, T. (2016). Different endosymbiotic interactions in two *Hydra* species reflect the evolutionary history of endosymbiosis. *Genome Biology and Evolution* 8, 2155–2163. https://doi.org/10.1093/gbe/evw142

Jablonka, E. & Lamb, M. J. (2008). Soft inheritance: challenging the modern synthesis. *Genetics and Molecular Biology* 31, 389–395. https://doi.org/10.1590/S1415-47572008000300001

Jacob, F. (1977). Evolution and tinkering. *Science* 196, 1161–1166. https://www.science.org/doi/10.1126/science.860134

Jaffe, L. F. (1966). Electrical currents through the developing *Fucas* egg. *Proceedings of the National Academy of Sciences USA* 56, 1102–1109. https://doi.org/10.1073/pnas.56.4.1102

Jaffe, L. F. & Nucitelli, R. (1977). Electrical controls of development. *Annual Review of Biophysics and Bioengineering* 6, 445–476. https://doi.org/10.1146/annurev.bb.06.060177.002305

Jain, A. & Zoncu, R. (2022). Organelle transporters and inter-organelle communication as drivers of metabolic regulation and cellular homeostasis. *Molecular Metabolism* 60, 101481. https://doi.org/10.1016/j.molmet.2022.101481

Jennings, H. S. (1906). *The Behavior of the Lower Organisms*. New York: Columbia University Press.

Jersáková, J., Johnson, S. D., & Kindlmann, P. (2006). Mechanisms and evolution of deceptive pollination in orchids. *Biological Reviews of Cambridge Philosophical Society* 81, 219–235. https://doi.org/10.1017/S1464793105006986

Jiménez-Marín, B. & Olson, B. J. S. C. (2022). The curious case of multicellularity in the Volvocine algae. *Frontiers in Genetics* 13, 787665. https://doi.org/10.3389/fgene.2022.787665

Johansson, J. S. (2003). Noninactivating tandem pore domain potassium channels as attractive targets for general anesthetics. *Anesthesia & Analgesia* 96, 1248–1250. https://doi.org/10.1213/01.ANE.0000058847.84859.54

Jones, D. P. & Sies, H. (2015). The redox code. *Antioxidant & Redox Signaling* 23, 734–746. https://doi.org/10.1089/ars.2015.6247

Jones, R. C. (2013). Science, sentience, and animal welfare. *Biological Philosophy* 28, 1–30. https://doi.org/10.1007/s10539-012-9351-1

Jorgensen, R. (1993). The origin of land plants: a union of alga and fungus advanced by flavonoids? *Biosystems* 31, 193–207. https://doi.org/10.1016/0303-2647(93)90049-i

Joukov, V. & De Nicolo, A. (2019). The centrosome and the primary cilium: the Yin and Yang of a hybrid organelle. *Cells* 8, 701. https://doi.org/10.3390/cells8070701

Joyce, G. F. & Szostak, J. W. (2018). Protocells and RNA self-replication. *Cold Spring Harbour Perspectives in Biology* 10, a034801. https://doi.org/10.1101/cshperspect.a034801

Juan, C. A., de la Lastra, J. M. P., Plou, F. J., & Pérez-Lebeña, E. (2021). The chemistry of reactive oxygen species (ROS) revisited: outlining their role in biological macromolecules (DNA, lipids and proteins) and induced pathologies. *International Journal of Molecular Science* 22, 4642. https://doi.org/10.3390/ijms22094642

Jung, J., Kim, J. S., Taffner, J., Berg, G., & Ryu, C. M. (2020). Archaea, tiny helpers of land plants. *Computational and Structural Biotechnology Journal* 18, 2494–2500. https://doi.org/10.1016/j.csbj.2020.09.005

Jung, S., Zimin, P. I., Woods, C. B., Kayser, E. B., Haddad, D., Reczek, C. R., et al. (2022). Isoflurane inhibition of endocytosis is an anesthetic mechanism of action. *Current Biology* 32, 3016–3032. https://doi.org/10.1016/j.cub.2022.05.03

Just, E. E. (1939). *The Biology of the Cell Surface*. Philadelphia, PA: P. Blakiston's Son & Co.

Kabadayi, C. & Osvath, M. (2017). Ravens parallel great apes in flexible planning for tool-use and bartering. *Science* 357, 202–204. https://doi.org/10.1126/science.aam8138

Kabir, S. R., Yokoyama, K., Mihashi, K., Kodama, T., & Suzuki, M. (2003). Hyper-mobile water is induced around actin filaments. *Biophysical Journal* 85, 3154–3161. https://doi.org/10.1016/S0006-3495(03)74733-X

Kaech, S., Brinkhaus, H., & Matus, A. (1999). Volatile anesthetics block actin-based motility in dendritic spines. *Proceedings of the National Academy of Sciences USA* 96, 10433–10437. https://doi.org/10.1073/pnas.96.18.10433

Kafatos, M. C. (2014). Physics and consciousness: quantum measurement, observation and experience. In: *White paper in the Workshop on the Frontiers of Consciousness Research*, 4 March, 1–14. National Academy of Science's Beckman Center at the University of California, Irvine.

Kagan, B. J., Kitchen, A. C., Tran, N. T., Habibollahi, F., Khajehnejad, M., Parker, B. J., et al. (2022). *In vitro* neurons learn and exhibit sentience when embodied in a simulated game-world. *Neuron* 110, 3952–3969. https://doi.org/10/1016/j.neuron2022.09.001

Kagenishi, T., Baluška, F., & Yokawa, K. (2022). Stress-induced ethanol affects endocytic vesicle recycling and F-actin organisation in Arabidopsis root apex cells. *Environmental and Experimental Botany* 205, 105123. https://doi.org/10.1016/j.envexpbot.2022.105123

Kaminski, J., Call, J., & Fischer, J. (2004). Word learning in a domestic dog: evidence for 'fast mapping'. *Science* 304, 1682–1683. https://doi.org/10.1126/science.1097859

Kauffman, S. A. (1993). *The Origins of Order: Self-Organization and Selection in Evolution.* New York: Oxford University Press.

Kauffman, S. A. (2000). *Investigations.* New York: Oxford University Press.

Kaur, D., Khaniya, U., Zhang, Y., & Gunner, M. R. (2021). Protein motifs for proton transfers that build the transmembrane proton gradient. *Frontiers in Chemistry* 9, 660954. https://doi.org/10.3389/fchem.2021.660954

Kawashima, T., Maruyama, D., Shagirov, M., Li, J., Hamamura, Y., Yelagandula, R., et al. (2014). Dynamic F-actin movement is essential for fertilization in *Arabidopsis thaliana. eLife* 3, e04501. https://doi.org/10.7554/eLife.04501

Keeling, P. J. (2010). The endosymbiotic origin, diversification and fate of plastids. *Philosophical Transactions of the Royal Society London B Biological Sciences* 365, 729–748. https://doi.org/10.1098/rstb.2009.0103

Kelly, J. R., Minuto, C., Cryan, J. F., Clarke, G., & Dinan, T. G. (2017). Cross talk: the microbiota and neurodevelopmental disorders. *Frontiers in Neuroscience* 11, 490.

Kelz, M. B. & Mashour, G. A. (2019). The biology of general anesthesia from paramecium to primate. *Current Biology* 29, R1199–R1210. https://doi.org/10.1016/j.cub.2019.09.07

Kempner, E. S. & Miller, J. H. (1968). The molecular biology of *Euglena gracilis.* V. Enzyme localization. *Experimental Cell Research* 51, 150–156. https://doi.org/10.1016/0014-827(68)90165-1

Kennedy, D. O. (2014). *Plants and the Human Brain.* Oxford: Oxford University Press.

Kessler, S. A. & Grossniklaus, U. (2011). She's the boss: signaling in pollen tube reception. *Current Opinion in Plant Biology* 14, 622–627. https://doi.org/10.1016/j.pbi.2011.07.012

Kessler, D., Diezel, C., & Baldwin, I. T. (2010). Changing pollinators as a means of escaping herbivores. *Current Biology* 20, 237–242. https://doi.org/10.1016/j.cub.2009.11.071

Key, B. (2016). Why fish do not feel pain. *Animal Sentience* 3, 1. https://doi.org/10.51291/2377-7478.1011

Key, B., Arlinghaus, R., & Browman, H. I. (2016). Insects cannot tell us anything about subjective experience or the origin of consciousness. *Proceedings of the National Academy of Sciences USA* 113, E3813. https://doi.org/10.1073/pnas.1606835113

Khanal, S., Leung, M. R., Royfman, A., Fishman, E. L., Saltzma, B., Bloomfield-Gadêlha, H., et al. (2021). A dynamic basal complex modulates mammalian sperm movement. *Nature Communications* 12, 3808. https://doi.org/10.1038/s41467-021-24011-0

Khrennikov, A. (2007). Can quantum information be processed by macroscopic systems? *Quantum Inf Process* 6, 401–429. https://link.springer.com/article/10.1007/s11128-007-0060-7

Kim, G. H., Klotchkova, T. A., & Kang, Y. M. (2014). Life without a cell membrane: regeneration of protoplasts from disintegrated cells of the marine green alga *Bryopsis plumosa. Journal of Cell Science* 114, 2009–2014. https://doi.org/10.1242/jcs.114.11.2009

Kim, H., Lee, H., Lee, G., Kim, H., & Cho, M. (2012). Hofmeister anionic effects on hydration electric fields around water and peptide. *Journal of Chemical Physics* 136, 124501. https://doi.org/10.1063/1.3694036

Kim, S., Coukos, R., Gao, F., & Krainc, D. (2022). Dysregulation of organelle membrane contact sites in neurological diseases. *Neuron* 110, 2386–2408. https://doi.org/10.1016/j.neuron.2022.04.020

Kinosita, K., Jr., Yasuda, R., Noji, H., & Adachi, K. (2000). A rotary molecular motor that can work at near 100% efficiency. *Philosophical Transaction of Royal Society London B Biological Sciences* 355, 473–489. https://doi.org/10.1098/rstb.2000.0589

Kirk, D. L. (2005). A twelve-step program for evolving multicellularity and a division of labor. *Bioessays* 2005, 27, 299–310. https://doi.org/10.1002/bies.20197

Kishikawa, J. I., Ibuki, T., Nakamura, S., Nakanishi, A., Minamino, T., Miyata, T., et al. (2013). Common evolutionary origin for the rotor domain of rotary ATPases and flagellar protein export apparatus. *PLoS One* 8, e64695. https://doi.org/10.1371/journal.pone.0064695

Klecker, T., Böckler, S., & Westermann, B. (2014). Making connections: interorganelle contacts orchestrate mitochondrial behavior. *Trends in Cell Biology* 24, 537–545. https://doi.org/10.1016/j.tcb.2014.04.004

Klein, C. & Barron, A. B. (2016). Insects have the capacity for subjective experience. *Animal Sentience* 9, 1.

Klier, P. E. Z., Gest, A. M. M., Martin, J. G., Roo, R., Navarro, M. X., Lesiak, L., et al. (2022). Bioorthogonal, fluorogenic targeting of voltage-sensitive fluorophores for visualizing membrane potential dynamics in cellular organelles. *Journal of the American Chemical Society* 144, 12138–12146. https://doi.org/10.1021/jacs.2c02664

Klink, V. P. & Wolniak, S. M. (2001). Centrin is necessary for the formation of the motile apparatus in spermatids of Marsilea. *Molecular Biology of the Cell* 12, 761–776. https://doi.org/10.1091/mbc.12.3.761

Klosin, A., Casas, E., Hidalgo-Carcedo, C., Vavouri, T., & Lehner, B. (2017). Transgenerational transmission of environmental information in *C. elegans*. *Science* 356, 320–323. https://doi.org/10.1126/science.aah6412

Knapp, B. D. & Huang, K. C. (2018). Translating the physical code of life. *Cell* 174, 253–255. https://doi.org/10.1016/j.cell.2018.06.050

Knoll, A. H., Bergmann, K. D., & Strauss, J. V. (2016). Life: the first two billion years. *Proceedings of the Royal Society B—Biological Sciences* 371, 20150493. https://doi.org/10.1098/rstb.2015.0493

Ko, M., Quiñones-Hinojosa, A., & Rao, R. (2020). Emerging links between endosomal pH and cancer. *Cancer and Metastasis Reviews* 39, 519–534. https://doi.org/10.1007/s10555-020-09870-1

Koch, C. (2018). What is consciousness? *Nature* 557, S9–S12. https://doi.org/10.1038/d41586-018-05097-x

Kodama, Y. & Fujishima, M. (2010). Secondary symbiosis between *Paramecium* and *Chlorella* cells. *International Review of Cell and Molecular Biology* 279, 33–77. https://doi.org/10.1016/S1937-6448(10)79002-X

Kodama, Y. & Fujishima, M. (2022). Endosymbiotic *Chlorella variabilis* reduces mitochondrial number in the ciliate *Paramecium bursaria*. *Scientific Reports* 12, 8216. https://doi.org/10.1038/s41598-022-12496-8

Kolluru, G. K., Shen, X., & Kevil, C. G. (2020). Reactive sulfur species. A new redox player in cardiovascular pathophysiology. *Arteriosclerosis, Thrombosis and Vascular Biology* 40, 874–884. https://doi.org/10.1161/ATVBAHA.120.314084

Kompanichenko, V. (2012). Inversion concept of the origin of life. *Origins of Life and Evolution of Biospheres* 42, 153–178. https://doi.org/10.1007/s11084-012-9279-0

Kondev, J. (2014). Bacterial decision making. *Physics Today* 67, 31–37.

Kondziella, D. (2020). The neurology of death and the dying brain: a pictorial essay. *Frontiers in Neurology* 11, 736. https://doi.org/10.3389/fneur.2020.00736

Koonin, E. V. (2007). The cosmological model of eternal inflation and the transition from chance to biological evolution in the history of life. *Biology Direct* 2(1), 1–21.

Koonin, E. V., Krupovic, M., Ishino, S., & Ishino, Y. (2020). The replication machinery of LUCA: common origin of DNA replication and transcription. *BMC Biology* 18, 61. https://doi.org/10.1186/s12915-020-00800-9

Koonin, E. V. & Wolf, Y. I. (2012). Evolution of microbes and viruses: a paradigm shift in evolutionary biology? *Frontiers in Cellular and Infection Microbiology* 2, 119. https://doi.org/10.3389/fcimb.2012.00119

Korenaga, J. (2021). Was there land on the early Earth? *Life* 11, 1142. https://doi.org/10.3390/life11111142

Koseska, A. & Bastiaens, P. I. (2017). Cell signaling as a cognitive process. *EMBO Journal* 36, 568–582. https://doi.org/10.15252/embj.201695383

Kostyrka, G. (2016). What roles for viruses in origin of life scenarios? *Studies in History and Philosophy of Science Part C: Studies in History and Philosophy of Biological and Biomedical Sciences*, 59, 135–144.

Koumandou, V. L., Wickstead, B., Ginger, M. L., van der Giezen, M., Dacks, J. B., & Field, M. C. (2013). Molecular paleontology and complexity in the last eukaryotic common ancestor. *Critical Reviews in Biochemistry and Molecular Biology* 48, 373–396. https://doi.org/10.3109/10409238.2013.821444

Kováč, L. (2006). Life, chemistry and cognition: conceiving life as knowledge embodied in sentient chemical systems might provide new insights into the nature of cognition. *EMBO Reports* 7, 562–566. https://doi.org/10.1038/sj.embor.7400717

Kováč, L. (2007). Information and knowledge in biology: time for reappraisal. *Plant Signaling & Behavior* 2, 65–73. https://doi.org/10.4161/psb.2.2.4113

Kováč, L. (2008). Bioenergetics: a key to brain and mind. *Communicative & Integrative Biology* 1, 114–122. https://doi.org/10.4161/cib.1.1.6670

Kovács, A. F., Fekete, N., Turiák, L., Ács, A., Kőhidai, L., Buzás, E. I., & Pállinger, E. (2019). Unravelling the role of trophoblastic-derived extracellular vesicles in regulatory T Cell differentiation. *International Journal of Molecular Sciences* 20, 3457. https://doi.org/10.3390/ijms20143457

Kowacz, M. & Pollack, G. H. (2021). Cells in new light: ion concentration, voltage, and pressure gradients across a hydrogel membrane. *ACS Omega* 5, 21024–21031. https://doi.org/10.1021/acsomega.0c02595

Kramar, M. & Alim, K. (2021). Encoding memory in tube diameter hierarchy of living flow network. *Proceedings of the National Academy of Sciences USA* 118, e2007815118

Kramer, B. A., Sarabia Del Castillo, J., & Pelkmans, L. (2022). Multimodal perception links cellular state to decision-making in single cells. *Science* 377, 642–648. https://doi.org/10.1126/science.abf4062

Krikorian, A. D. (1975). Were the opium poppy and opium known in the ancient near east? *Journal of the History of Biology* 8, 95–114. PMID: 11609871.

Kreimer, G. (2009). The green algal eyespot apparatus: a primordial visual system and more? *Current Genetics* 55, 19–43. https://doi.org/10.1007/s00294-008-0224-8

Kühlbrandt, W. (2019). Structure and mechanisms of F-type ATP synthases. *Annual Reviews of Biochemistry* 88, 515–549. https://doi.org/10.1146/annurev-biochem-013118-110903

Kühlbrandt, W. & Davies, K. M. (2016). Rotary ATPases: a new twist to an ancient machine. *Trends in Biochemical Sciences* 41, 106–116. https://doi.org/10.1016/j.tibs.2015.10.006

Kühn, M. J., Talà, L., Inclan, Y. F., Patino, R., Pierrat, X., Vos, I., et al. (2021). Mechanotaxis directs *Pseudomonas aeruginosa* twitching motility. *Proceedings of the National Academy of Sciences USA* 118, e2101759118. https://doi.org/10.1073/pnas.2101759118

Kühne, W. (1864). *Untersuchungen über das Protoplasma und die Contractilität*. Leipzig: W. Engelmann.

Kundacina, N., Shi, M., & Pollack, G. H. (2016). Effect of local and general anesthetics on interfacial water. *PLoS One* 11, e0152127. https://doi.org/10.1371/journal.pone.015212

Kunkel, T. A. (2009, January). Evolving views of DNA replication (in) fidelity. In *Cold Spring Harbor symposia on quantitative biology* (Vol. 74, pp. 91–101). Cold Spring Harbor Laboratory Press.

Kwapiszewska, K., Szczepański, K., Kalwarczyk, T., Michalska, B., Patalas-Krawczyk, P., Szymański, J., et al. (2020). Nanoscale viscosity of cytoplasm is conserved in human cell lines. *Journal of Physical Chemistry Letters* 11, 6914–6920. https://doi.org/10.1021/acs.jpclett.0c01748

Laage, D., Elsaesser, T., & Hynes, J. T. (2017). Structure and ultrafast dynamics of biomolecular hydration shells. *Structural Dynamics* 4, 044018. https://doi.org/10.1063/1.4981019

LaBella, F. S., Stein, D., & Queen, G. (1999). The site of general anesthesia and cytochrome P450 monooxygenases: occupation of the enzyme heme pocket by xenon and nitrous oxide. *European Journal of Pharmacology* 381, R1–R3. https://doi.org/10.1016/s0014-2999(99)00553-1

Laetz, E. M., Moris, V. C., Moritz, L., Haubrich, A. N., & Wägele, H. (2017). Photosynthate accumulation in solar-powered sea slugs—starving slugs survive due to accumulated starch reserves. *Frontiers in Zoology* 14, 4. https://doi.org/10.1186/s12983-016-0186-5

Laflamme, G. & Mekhail, K. (2020). Biomolecular condensates as arbiters of biochemical reactions inside the nucleus. *Communications Biology* 3, 773. https://doi.org/10.1038/s42003-020-01517-9

La Fortezza, M. & Velicer, G. J. (2021). Social selection within aggregative multicellular development drives morphological evolution. *Proceedings of the Royal Society B Biological Sciences* 288, 20211522. https://doi.org/10.1098/rspb.2021.1522

Laiz, L. & Taiz, L. (2017). *Flora Unveiled: The Discovery and Denial of Sex in Plants*. Oxford: Oxford University Press

Lancet, D., Zidovetzki, R., & Markovitch, O. (2018). Systems protobiology: origin of life in lipid catalytic networks. *Journal of the Royal Society Interface* 15, 20180159. https://doi.org/10.1098/rsif.2018.0159

Lane, N. (2015). *The Vital Question. Energy, Evolution, and the Origins of Complex Life*. London: W. W. Norton & Company Ltd.

Lane, N. (2017). Proton gradients at the origin of life. *BioEssays* 39, 1600217. https://doi.org/10.1002/bies.201600217

Lane, N. (2022). *Transformer: The Deep Chemistry of Life and Death*. London: Profile Books Ltd.

Langhorst, M. F., Reuter, A., & Stuermer, C. A. (2007). Scaffolding microdomains and beyond: the function of reggie/flotillin proteins. *Cellular and Molecular Life Sciences* 62, 2228–2240. https://doi.org/10.1007/s00018-005-5166-4

Lanna, E. & Klautau, M. (2010). Oogenesis and spermatogenesis in *Paraleucilla magna* (Porifera, Calcarea). *Zoomorphology* 129, 249–261. https://doi.org/10.1007/s00435-010-0117-5

Larson, E. J. (2021). *The Myth of Artificial Intelligence: Why Computers Can't Think the Way We Do*. The Belknap Press of Harvard University Press.

Lebiedzinska, L., Szabadkai, G., Jones, A. W. E., Duszynski, J., & Wieckowski, M. R. (2009). Interactions between the endoplasmic reticulum, mitochondria, plasma membrane and other subcellular organelles. *International Journal of Biochemistry & Cell Biology* 41, 1805–1816. https://doi.org/10.1016/j.biocel.2009.02.017

LeDoux, J. (2019). *The Deep History of Ourselves: The Four-Billion Year Story of How We Got Conscious Brains*. New York: Penguin Books.

Lee, K. & Seo, P. J. (2022). Wound-induced systemic responses and their coordination by electrical signals. *Frontiers in Plant Sciences* 13, 880680. https://doi.org/10.3389/fpls.2022.880680

Lefièvre, L., Machado-Oliveira, G., Ford, C., Kirkman-Brown, J., Barratt, C., & Publicover, S. (2009). Communication between female tract and sperm. Saying NO• when you mean yes. *Communicative & Integrative Biology* 2, 82–85. https://doi.org/10.4161/cib.7502

Lenton, T. M. & Latour, B. (2018). Gaia 2.0. *Science* 361, 1066–1068. https://doi.org/10.1126/science.aau0427

Lever, M. J., Miller, K. W., Paton, W. D., & Smith, E. B. (1971). Pressure reversal of anesthesia. *Nature* 231, 368–371. https://doi.org/10.1038/231368a0

Levin, M. (2014). Endogenous bioelectrical networks store non-genetic patterning information during development and regeneration. *Journal of Physiology* 592, 2295–2305. https://doi.org/10.1113/jphysiol.2014.271940

Levin, M. (2019). The computational boundary of a 'self': developmental bioelectricity drives multicellularity and scale-free cognition. *Frontiers in Psychology* 10, 2688. https://doi.org/10.3389/fpsyg.2019.02688

Levin, M. (2021). Bioelectric signaling: reprogrammable circuits underlying embryogenesis, regeneration, and cancer. *Cell* 184, 1971–1989. https://doi.org/10.1016/j.cell.2021.02.034

Levin, M. (2022). Generalizing frameworks for sentience beyond natural species: commentary on Crump, et al. on decapod sentience. *Animal Sentience* 32, 429.

Levin, M. & Martyniuk, C. J. (2018). The bioelectric code: an ancient computational medium for dynamic control of growth and form. *Biosystems* 164, 76–93. https://doi.org/10.1016/j.biosystems.2017.08.009

Levin, M., Pezzulo, G., & Finkelstein, J. M. (2017). Endogenous bioelectric signaling networks: exploiting voltage gradients for control of growth and form. *Annual Review of Biomedical Engineering* 19, 353–387. https://doi.org/10.1146/annurev-bioeng-071114-040647

Lev-Yadun, S. (2001). Intrusive growth—the plant analog of dendrite and axon growth in animals. *New Phytologist* 150, 508–512. https://doi.org/10.1046/j.1469-8137.2001.00143.x

Li, J., Liu, X., Abdelmohsen, L. K. E. A., Williams, D. S., & Huang, X. (2019). Spatial organization in proteinaceous membrane-stabilized coacervate protocells. *Small* 15, 1902893. https://doi.org/10.1002/smll.201902893

Li, L., Dou, N., Zhang, H., & Wu, C. (2021). The versatile GABA in plants. *Plant Signaling & Behavior* 16, 1862565. https://doi.org/10.1080/15592324.2020

Li, W. & Jiang, H. (2022). Nuclear protein condensates and their properties in regulation of gene expression. *Journal of Molecular Biology* 434, 167151. https://doi.org/10.1016/j.jmb.2021.167151

Lieff, J. (2020). *The Secret Language of Cells*. Dallas, TX: Ben Bella Books.

Lin, E. C. & Cantiello, H. F. (1993). A novel method to study the electrodynamic behavior of actin filaments. Evidence for cable-like properties of actin. *Biophysical Journal* 65, 1371–1378. https://doi.org/10.1016/S0006-3495(93)81188-3

Lindås, A. C., Valegård, K., & Ettema, T. J. G. (2017). Archaeal actin-family filament systems. *Subcellular Biochemistry* 84, 379–392. https://doi.org/10.1007/978-3-319-53047-5_13

Lindeman, M., & Saher, M. (2007). Vitalism, purpose and superstition. *British Journal of Psychology* 98, 33–44. https://doi.org/10.1348/000712606x101808

Line, M. A. (2007). Panspermia in the context of the timing of the origin of life and microbial phylogeny. *International Journal of Astrobiology* 6, 249–254. https://doi.org/10.1017/S1473550407003813

Ling, D. (2007). Nano-protoplasm: the ultimate unit of life. *Physiological Chemistry and Physics and Medical NMR* 39, 111–234. PMID: 19256352.

Lingwood, D. & Simons, K. (2010). Lipid rafts as a membrane-organizing principle. *Science* 327, 46–50. https://doi.org/10.1126/science.1174621

Liu, J., Cvirkaite-Krupovic, V., Commere, P. H., Yang, Y., Zhou, F., Forterre, P., et al. (2021). Archaeal extracellular vesicles are produced in an ESCRT-dependent manner and promote gene transfer and nutrient cycling in extreme environments. *ISME Journal* 15, 2892–2905. https://doi.org/10.1038/s41396-021-00984-0

Liu, J., Martinez-Corral, R., Prindle, A., Lee, D.-Y. D., Larkin, J., Gabalda-Sagarra, M., et al. (2017). Coupling between distant biofilms and emergence of nutrient time-sharing. *Science* 356, 638–642. https://doi.org/10.1126/science.aah4204

Liu, S., Jiao, J., Lu, T. J., Xu, F., Pickard, B. G., & Genin, G. M. (2017). Arabidopsis leaf trichomes as acoustic antennae. *Biophysical Journal* 113, 2068–2076. https://doi.org/10.1016/j.bpj.2017.07.035.46

Lloyd, S. (2002). Computational capacity of the universe. *Physical Review Letters* 88, 237901. https://doi.org/10.1103/PhysRevLett.88.237901

Lombard, J. (2014). Once upon a time the cell membranes: 175 years of cell boundary research. *Biology Direct* 9, 32. https://doi.org/10.1186/s13062-014-0032-7

Lombardi, O., Fortin, S., & Vanni, L. (2015). A pluralist view about information. *Philosophy of Science* 82, 1248–1259.

Lombardi, O., Holik, F., & Vanni, L. (2016). What is Shannon information? *Synthese* 193, 1983–2012. https://doi.org/10.1007/s11229-015-0824-z

Loudon, A. S. I. (2012). Circadian biology: a 2.5 billion year old clock. *Current Biology* 22, R570–R571. https://doi.org/10.1016/j.cub.2012.06.023

Lu, Y. & Zhong, D. (2019). Understanding short-range electron-transfer dynamics in proteins. *Journal of Physical Chemistry Letters* 10, 346–351. https://doi.org/10.1021/acs.jpclett.8b0374

Luby-Phelps, K. (2000). Cytoarchitecture and physical properties of cytoplasm: volume, viscosity, diffusion, intracellular surface area. *International Review of Cytology* 192, 189–221. https://doi.org/10.1016/s0074-7696(08)60527-6

Luby-Phelps, K. (2013). The physical chemistry of cytoplasm and its influence on cell function: an update. *Molecular Biology of the Cell* 24, 2593–2596. https://doi.org/10.1091/mbc.E12-08-0617

Luckhardt, A. & Carter, J. B. (1923). Ethylene as a gas anesthetic. *Anesthesia & Anelgesia* 2, 221–229. https://doi.org/10.1001/jama.1923.02640470018007

Luczak, A., & Kubo, Y. (2021). Predictive neuronal adaptation as a basis for consciousness. *Frontiers in Systems Neuroscience* 15, 767461. https://doi.org/10.3389/fnsys.2021.767461

Luczak, A., McNaughton, B. L., & Kubo, Y. (2022). Neurons learn by predicting future activity. *Nature Machine Intelligence* 4, 62–72. https://doi.org/10.1038/s42256-021-00430-y

Lund, E. J. & Kenyon, W. A. (1927). Relation between continuous bio-electric currents and cell respiration. I. Electric correlation potentials in growing root tip. *Journal of Experimental Zoology* 48, 333–357. https://doi.org/10.1002/jez.140048020

Lynch, J. B. & Hsiao, E. Y. (2019). Microbiomes as sources of emergent host phenotypes. *Science* 365, 1405–1409. https://doi.org/10.1126/science.aay0240

Lyon, P. (2015). The cognitive cell: bacterial behavior reconsidered. *Frontiers in Microbiology* 6, 264. https://doi.org/10.3389/fmicb.2015.00264

Ma, C., Wang, C., Luo, D., Yan, L., Yang, W., Li, N., & Gao, N. (2022). Structural insights into the membrane microdomain organization by SPFH family proteins. *Cell Research* 32, 176–189. https://doi.org/10.1038/s41422-021-00598-3

MacDonald, A. G. (1994). A short history of fires and explosions caused by anaesthetic agents. *British Journal of Anesthesia* 72, 710–722. https://doi.org/10.1093/bja/72.6.710

Machado-Oliveira, G., Lefièvre, L., Ford, C., Herrero, M. B., Barratt, C., Connolly, T. J., Nash, K., Morales-Garcia, A., Kirkman-Brown, J., & Publicover, S. (2008). Mobilisation of Ca2+ stores and flagellar regulation in human sperm by S-nitrosylation: a role for NO synthesised in the female reproductive tract. *Development* 135, 3677–3686. https://doi.org/10.1242/dev.024521

Macnab, R. M. & Koshland, D. E., Jr. (1972). Gradient-sensing mechanisms in bacterial chemotaxis. *Proceedings of the National Academy of Sciences USA* 69, 2509–2512. https://doi.org/10.1073/pnas.69.9.2509

Macphail, E. (1998). *The Evolution of Consciousness*. Oxford: Oxford University Press.

Mahnert, A., Blohs, M., Pausan, M. R., & Moissl-Eichinger, C. (2018). The human archaeome: methodological pitfalls and knowledge gaps. *Emerging Topics in Life Sciences* 2, 469–482. https://doi.org/10.1042/ETLS20180037

Mak, D. O., Vais, H., Cheung, K. H., & Foskett, J. K. (2013). Isolating nuclei from cultured cells for patch-clamp electrophysiology of intracellular Ca^{2+} channels. *Cold Spring Harbor Protocols* 880–884. https://doi.org/10.1101/pdb.prot073056

Mallatt, J. & Feinberg, T. E. (2017). Consciousness is not inherent in but emergent from life. Commentary on Reber on Origins of Mind. *Animal Sentience* 11, 15.

Manchester, L. C., Coto-Montes, A., Boga, J. A., Andersen, L. P., Zhou, Z., Galano, A., et al. (2015). Melatonin: an ancient molecule that makes oxygen metabolically tolerable. *Journal of Pineal Research* 59, 403–419. https://doi.org/10.1111/jpi.12267

Manna, S., Ghanty, C., Baindara, P., Barik, T. K., & Mandal, S. M. (2020). Electrochemical communication in biofilm of bacterial community. *Journal of Basic Microbiology* 60, 819–827. https://doi.org/10.1002/jobm.202000340

Manrique-Bedoya, S. & Marucho, M. (2022). Molecular structure study on the polyelectrolyte properties of actin filaments. *RSC Advances* 12, 6314–6327. https://doi.org/10.1039/d1ra09280d

Margulis, L. (2001). The conscious cell. *Annals of the New York Academy of Sciences* 929, 55–70. https://doi.org/10.1111/j.1749-6632.2001.tb05707.x

Margulis, L., Chapman, M., Guerrero, R., & Hall, J. (2006). The last eukaryotic common ancestor (LECA): acquisition of cytoskeletal motility from aerotolerant spirochetes in the Proterozoic eon. *Proceedings of the National Academy of Sciences USA* 103, 13080–13085. https://doi.org/10.1073/pnas.0604985103

Marijuán, P. C. & Navarro, J. (2022). The biological information flow: from cell theory to a new evolutionary synthesis. *Biosystems* 213, 104631. https://doi.org/10.1016/j.biosystems.2022.104631

Marijuán, P. C., Navarro, J., & del Moral, R. (2015). How the living is in the world: an inquiry into the informational choreographies of life. *Progress in Biophysics and Molecular Biology* 31, 469–480. https://doi.org/10.1016/j.pbiomolbio.2015.07.002

Markoš, A. (1995). The ontogeny of Gaia: the role of microorganisms in planetary information network. *Journal of Theoretical Biology* 176, 175–180. https://doi.org/10.1006/jtbi.1995.0186

Markoš, A. (2002). *Readers of the Book of Life: Contextualizing Developmental Evolutionary Biology*. New York: Oxford University Press.

Marsh, G. (1928). Relation between continuous bio-electric currents and cell respiration. IV. The origin of electric polarity in the onion root. *Journal of Experimental Zoology* 51, 309–325. https://doi.org/10.1002/jez.140051030

Martin, W., Baross, J., Kelley, D., & Russell, M. J. (2008). Hydrothermal vents and the origin of life. *Nature Reviews Microbiology* 6, 805–814. https://doi.org/10.1038/nrmicro1991

Martinez-Corral, R., Liu, J., Prindle, A., Süel, G. M., & Garcia-Ojalvo, J. (2019). Metabolic basis of brain-like electrical signalling in bacterial communities. *PLoS Computational Biology* 15, e1007508. https://doi.org/10.1371/journal.pcbi.1007508

Masi, M. (2022). Vitalism and cognition in a conscious universe. *Communicative & Integrative Biology* 15, 121–136. https://doi.org/10.1080/19420889.2022.2071102

Masi, E., Ciszak, M., Stefano, G., Renna, L., Azzarello, E., Pandolfi, C., et al. (2009). Spatiotemporal dynamics of the electrical network activity in the root apex. *Proceedings of the National Academy of Sciences USA* 106, 4048–4053. https://doi.org/10.1073/pnas.0804640106

Matamalo, E., Castillo, C., Vivar, J. P., Rojas, P. A., & Brauchi, S. E. (2021). Imaging the electrical activity of organelles in living cells. *Communicative Biology* 4, 389. https://doi.org/10.1038/s42003-021-01916-6

Mathis, R. & Ackermann, M. (2016). Response of single bacterial cells to stress gives rise to complex history dependence at the population level. *Proceedings of the National Academy of Sciences USA* 113, 4334–4229. https://doi.org/10.1073/pnas.1511509113

Mathur, J., Kroeker, O. F., Lobbezoo, M., & Mathur, N. (2022). The ER is a common mediator for the behavior and interactions of other organelles. *Frontiers of Plant Science* 13, 846970. https://doi.org/10.3389/fpls.2022.846970

Matos, B., Publicover, S. J., Castro, L. F. C., Esteves, P. J., & Fardilha, M. (2021). Brain and testis: more alike than previously thought? *Open Biology* 11, 200322. https://doi.org/10.1098/rsob.200322

Mattiroli, F., Bhattacharyya, S., Dyer, P. N., White, A. E., Sandman, K., Burkhart, B. W., et al. (2017). Structure of histone-based chromatin in Archaea. *Science* 357, 609–661.

Matzke, A. J. M., Lin, W. D., & Matzke, M. (2019). Evidence that ion-based signaling initiating at the cell surface can potentially influence chromatin dynamics and chromatin-bound proteins in the nucleus. *Frontiers in Plant Science* 10, 1267. https://doi.org/10.3389/fpls.2019.01267

Mayer, F., Wheatley, D., & Hoppert, M. (2006). Some properties of interfacial water: determinants for cell architecture and function? In: Pollack, G., Cameron, I. L., & Wheatley, D. N. (Eds.), *Water and the Cell*, pp. 253–271. Dortrecht: Springer Verlag.

Maynard Smith, J. (2000). The concept of information in biology. *Philosophy of Science* 67, 177–194.

Mazor, M., Brown, S., Ciaunica, A., Demertzi, A., Fahrenfort, J., Faivre, N., et al. (2022). The scientific study of consciousness cannot and should not be morally neutral. *Perspectives on Psychological Science*. First published online 28 September 2022. https://doi.org/10.1177/17456916221110222

Mazzarello, P. (1999). A unifying concept: the history of cell theory. *Nature Cell Biology* 1, E13–E15. https://doi.org/10.1038/8964

McElroy, J. S. (2014). Vavilovian mimicry: Nikolai Vavilov and his little-known impact on weed science. *Weed Science* 62, 207–216. https://doi.org/10.1614/WS-D-13-00122.1

McFadden, G. I. (2014). Origin and evolution of plastids and photosynthesis in eukaryotes. *Cold Spring Harbor Perspectives in Biology* 6, a016105. https://doi.org/10.1101/cshperspect.a016105

McGlynn, S. E., Chadwick, G. L., Kempes, C. P., & Orphan, V. J. (2015). Single cell activity reveals direct electron transfer in methanotrophic consortia. *Nature* 526, 531–535. https://doi.org/10.1038/nature15512

McMillan, H. M. & Kuehn, M. J. (2021). The extracellular vesicle generation paradox: a bacterial point of view. *EMBO Journal* 40, e108174. https://doi.org/10.15252/embj.2021108174

Meng, B. & Lever, A. M. (2021). The interplay between ESCRT and viral factors in the enveloped virus life cycle. *Viruses* 13, 324. https://doi.org/10.3390/v13020324

Merchant, S. S., Prochnik, S. E., Vallon, O., Harris, E. H., Karpowicz, S. J., Witman, G. B., et al. (2007). The *Chlamydomonas* genome reveals the evolution of key animal and plant functions. *Science* 318, 245–250. https://doi.org/10.1126/science.1143609

Merker, B. (2005). The liabilities of mobility: a selection pressure for the transition to consciousness in animal evolution. *Consciousness and Cognition* 14, 89–114. https://doi.org/10.1016/S1053-8100(03)00002-3

Merker, B. (2007). Consciousness without a cerebral cortex: a challenge for neuroscience and medicine. *Behavioral and Brain Sciences* 30, 63–81. https://doi.org/10.1017/S0140525X07000891

Mescher, M. C., Runyon, J. B., & De Moraes, C. M. (2006). Plant host finding by parasitic plants—a new perspective on plant to plant communication. *Plant Signaling & Behavior* 1, 284–286. https://doi.org/10.4161/psb.1.6.3562

Messori, C. (2019). The super-coherent state of biological water. *Open Access Library Journal* 6, e5236. https://doi.org/10.4236/oalib.1105236

Michaelian, K. (2022). Non-equilibrium thermodynamic foundations of the origin of life. *Foundations* 2, 308–337. https://doi.org/10.3390/foundations2010022

Michard, E., Lima, P. T., Borges, F., Silva, A. C., Portes, M. T., Carvalho, J. E., et al. (2011). Glutamate receptor-like genes form Ca^{2+} channels in pollen tubes and are regulated by pistil D-serine. *Science* 332, 434–437. https://doi.org/10.1126/science.1201101

Michieletto, D. & Marenda, M. (2022). Rheology and viscoelasticity of proteins and nucleic acids condensates. *JACS Au* 2, 1506–1521. https://doi.org/10.1021/jacsau.2c00055

Mikhalevich, I. & Powell, R. (2020). Minds without spines: evolutionarily inclusive animal ethics. *Animal Sentience* 29, 1. https://doi.org/10.51291/2377-7478.1527

Mikhailovsky, G. E. & Gordon, R. (2021). LUCA to LECA, the Lucacene: a model for the gigayear delay from the first prokaryote to eukaryogenesis. *Biosystems* 205, 104415. https://doi.org/10.1016/j.biosystems.2021.104415

Milev, N. B. & Reddy, A. B. (2015). Circadian redox oscillations and metabolism. *Trends in Endocrinology and Metabolism* 26, 430–437. https://doi.org/10.1016/j.tem.2015.05.012

Milev, N. B., Rhee, S. G., & Reddy, A. B. (2018). Cellular timekeeping: it's redox o'clock. *Cold Spring Harbour Perspectives in Biology* 10, a027698. https://doi.org/10.1101/cshperspect.a027698

Miller, E., Morel, A., Saso, L., & Saluk, J. (2015). Melatonin redox activity. Its potential clinical applications in neurodegenerative disorders. *Current Topics in Medicinal Chemistry* 15, 163–169. https://doi.org/10.2174/1568026615666141209160556

Miller, J. H., Jr., Rajapakshe, K. I., Infante, H. L., & Claycomb, J. R. (2013). Electric field driven torque in ATP synthase. *PLoS One* 8, e74978. https://doi.org/10.1371/journal.pone.0074978

Miller, W. B., Jr. (2013). *The Microcosm Within: Evolution and Extinction in the Hologenome.* Boca Raton, FL: Universal-Publishers.

Miller, W. B., Jr. (2016a). Cognition, information fields and hologenomic entanglement: evolution in light and shadow. *Biology* 5, 21. https://doi.org/10.3390/biology5020021

Miller, W. B., Jr. (2016b). The eukaryotic microbiome: origins and implications for fetal and neonatal life. *Frontiers in Pediatrics* 4, 96. https://doi.org/10.3389/fped.2016.00096

Miller, W. B., Jr. (2018). Biological information systems: evolution as cognition-based information management. *Progress in Biophysics and Molecular Biology* 134, 1–26. https://doi.org/10.1016/j.pbiomolbio.2017.11.005

Miller, W. B. (2023). A scale-free universal relational information matrix (N-space) reconciles the information problem: N-space as the fabric of reality. *Communicative & Integrative Biology* 16(1), 2193006.

Miller, W. B., Jr., Baluška, F., & Torday, J. S. (2020a). Cellular senomic measurements in cognition-based evolution. *Progress in Biophysics and Molecular Biology* 156, 20–33. https://doi.org/10.1016/j.pbiomolbio.2020.07.002

Miller, W. B., Jr., Enguita, F. J., & Leitão, A. L. (2021). Non-random genome editing and natural cellular engineering in cognition-based evolution. *Cells* 10, 1125. https://doi.org/10.3390/cells10051125

Miller, W. B., Jr., Reber, A. S., Marshall, P., & Baluška, F. (2023). Cellular and viral natural engineering in cognition-based evolution. *Communicative & Integrative Biology* 16(1), 2196145.

Miller, W. B., Jr., & Torday, J. S. (2018). Four domains: the fundamental unicell and post-Darwinian cognition-based evolution. *Progress in Biophysics and Molecular Biology* 140, 49–73. https://doi.org/10.1016/j.pbiomolbio.2018.04.006

Miller, W. B., Torday, J. S., & Baluška, F. (2019). Biological evolution as the defense of self. *Progress in Biophysics and Molecular Biology* 142, 54–74. https://doi.org/10.1016/j.pbiomolbio.2018.10.002

Miller, W. B., Jr., Torday, J. S., & Baluška, F. (2020b). The N-space Episenome unifies cellular information space-time within cognition-based evolution. *Progress in Biophysics and Molecular Biology* 150, 112–139. https://doi.org/10.1016/j.pbiomolbio.2019.08.006

Mills, D. B. (2020). The origin of phagocytosis in Earth history. *Interface Focus* 10, 20200019. https://doi.org/10.1098/rsfs.2020.0019

Mills, D. B., Boyle, R. A., Daines, S. J., Sperling, E. A., Pisani, D., Donoghue, P. C. J., et al. (2022). Eukaryogenesis and oxygen in Earth history. *Nature Ecology & Evolution* 6, 520–532. https://doi.org/10.1038/s41559-022-01733-y

Milshteyn, D., Cooper, G., & Deamer, D. (2019). Chemiosmotic energy for primitive cellular life: proton gradients are generated across lipid membranes by redox reactions coupled to meteoritic quinones. *Scientific Reports* 9, 12447. https://doi.org/10.1038/s41598-019-48328-5

Mitchell, A., Romano, G. H., Groisman, B., Yona, A., Dekel, E., Kupiec, M., et al. (2009). Adaptive prediction of environmental changes by microorganisms. *Nature* 460, 220–224. https://doi.org/10.1038/nature08112

Mitchell, P. D. (1961). Coupling of phosphorylation to electron and hydrogen transfer by a chemi-osmotic type of mechanism. *Nature* 191, 144–148. https://doi.org/10.1038/191144a

Mittleman, B. E., Pott, S., Warland, S., Barr, K., Cuevas, C., & Gilad, Y. (2021). Divergence in alternative polyadenylation contributes to gene regulatory differences between humans and chimpanzees. *eLife* 10, e62548. https://doi.org/10.7554/eLife.62548

Miyakawa, S., Cleaves, H. J., & Miller, S. L. (2002). The cold origin of life: implications based on the hydrolytic stabilities of hydrogen cyanide and formamide. *Origins of Life and Evolution of the Biosphere* 32, 195–208. https://doi.org/10.1023/a:1016514305984

Miyamoto, Y., Muto, E., Mashimo, T., Iwane, A. H., Yoshiya, I., & Yanagida, T. (2000). Direct inhibition of microtubule-based kinesin motility by local anesthetics. *Biophysical Journal* 78, 940–949. https://doi.org/10.1016/S0006-3495(00)76651-3

Miyokawa, R., Hanada, M., Togawa, Y., Itoh, T. Q., Kobayakawa, Y., & Kusumi, J. (2022). Symbiont specificity differs among green hydra strains. *Royal Society Open Science* 9, 220789. https://doi.org/10.1098/rsos.220789

Mizraji, E. (2021). The biological Maxwell's demons: exploring ideas about the information processing in biological systems. *Theory in Biosciences* 140, 307–318. https://doi.org/10.1007/s12064-021-00354-6

Mladenov, V., Fotopoulos, V., Kaiserli, E., Karalija, E., Maury, S., Baranek, M., et al. (2021). Deciphering the epigenetic alphabet involved in transgenerational stress memory in crops. *International Journal of Molecular Sciences* 22, 7118. https://doi.org/10.3390/ijms22137118

Moi, D., Nishio, S., Li, X., Valansi, C., Langleib, M., Brukman, N. G., et al. (2022). Discovery of archaeal fusexins homologous to eukaryotic HAP2/GCS1 gamete fusion proteins. *Nature Communications* 13, 3880. https://doi.org/10.1038/s41467-022-31564-1

Mollenhauer, H. H. & Morré, D. J. (1978). Structural compartmentation of the cytosol: zones of exclusion, zones of adhesion, cytoskeletal and intercisternal elements. *Subcellular Biochemistry* 5, 327–359. https://doi.org/10.1007/978-1-4615-7942-7_7

Moran, J. (2022). Bringing inanimate matter to life. *Science* 377, 39. https://doi.org/10.1126/science.abp8953

Morowitz, H. J. (1978). Proton semiconductors and energy transduction in biological system. *American Journal of Physiology* 35, R99–R114. https://doi.org/10.1152/ajpregu.1978.235.3.R99

Morowitz, H. J., Heinz, B., & Deamer, D. W. (1988). The chemical logic of a minimum protocell. *Origins of Life and Evolution of the Biosphere* 18, 281–287. https://doi.org/10.1007/BF01804674

Morowitz, H. & Smith, E. (2007). Energy flow and the organization of life. *Complexity* 13, 51–59. https://doi.org/10.1002/cplx.20191

Morrow, I. C. & Parton, R. G. (2005). Flotillins and the PHB domain protein family: rafts, worms and anaesthetics. *Traffic* 6, 725–740. https://doi.org/10.1111/j.1600-0854.2005.00318.x

Mu, J., Zhuang, X., Wang, Q., Jiang, H., Deng, Z., Wang, B., et al. (2014). Interspecies communication between plant and mouse gut host cells through edible plant derived exosome-like nanoparticles. *Molecular Nutrition & Food Research* 58, 1561–1573. https://doi.org/10.1002/mnfr.201300729

Muchowska, K. B., Varma, S. J., & Moran, J. (2020). Nonenzymatic metabolic reactions and life's origins. *Chemical Reviews* 120, 7708–7744. https://doi.org/10.1021/acs.chemrev.0c00191

Mühlbauer, M. E., Gamiz-Hernandez, A. P., & Kaila, V. R. I. (2021). Functional dynamics of an ancient membrane-bound hydrogenase. *Journal of the American Chemical Society* 143, 20873–20883. https://doi.org/10.1021/jacs.1c09356

Mukherjee, S. (2022). *The Song of the Cell*. New York: Simon & Schuster, Inc.

Mulkidjanian, A. Y., Bychkov, A. Y., Dibrova, D. V., Galperin, M. Y., & Koonin, E. V. (2012). Origin of first cells at terrestrial, anoxic geothermal fields. *Proceedings of the National Academy of Sciences USA* 109, E821–E830. https://doi.org/10.1073/pnas.1117774109

Mulkidjanian, A. Y. & Galperin, M. Y. (2009). On the origin of life in the zinc world. 2. Validation of the hypothesis on the photosynthesizing zinc sulfide edifices as cradles of life on Earth. *Biology Direct* 4, 1–37. https://doi.org/10.1186/1745-6150-4-27

Muller, A. W. (2005). Thermosynthesis as energy source for the RNA World: a model for the bioenergetics of the origin of life. *Biosystems* 82, 93–102. https://doi.org/10.1016/j.biosystems.2005.06.003

Müller, J. J. A., Massen, J. J. M., Bugnyar, T., & Osvath, M. (2017). Ravens remember the nature of a single reciprocal interaction sequence over 2 days and even after a month. *Animal Behaviour* 128, 69–78. https://doi.org/10.1016/j.anbehav.2017.04.004

Müller, V., Lingl, A., Lewalter, K., & Fritz, M. (2005). ATP synthases with novel rotor subunits: new insights into structure, function and evolution of ATPases. *Journal of Bioenergetics and Biomembranes* 37, 455–460. https://doi.org/10.1007/s10863-005-9491-y

Munro, S. (2003). Lipid rafts: elusive or illusive? *Cell* 115, 377–388. https://doi.org/10.1016/s0092-8674(03)00882-1

Muscatine, L. & Lenhoff, H. M. (1963). Symbiosis: on the role of algae symbiotic with hydra. *Science* 142, 956–958. https://doi.org/10.1126/science.142.3594.956

Mussel, M. & Schneider, M. F. (2019). It sounds like an action potential: unification of electrical, chemical and mechanical aspects of acoustic pulses in lipids. *Journal of the Royal Society Interface* 16, 20180743. https://doi.org/10.1098/rsif.2018.0743

Mustard, J. A. (2020). Neuroactive nectar: compounds in nectar that interact with neurons. *Arthropod-Plant Interactions* 14, 151–159. https://doi.org/10.1007/s11829-020-09743-y

Myrum, C., Moreno-Castilla, P., & Rapp, P. R. (2022). 'Arc'-hitecture of normal cognitive aging. *Ageing Research Reviews* 80, 101678. https://doi.org/10.1016/j.arr.2022.101678

Nagel, T. (1974). What is it like to be a bat? *The Philosophical Review* 83, 435–450. https://doi.org/10.2307/2183914

Nakanishi-Matsui, M., Skiya, M., Nakamoto, R. K., & Futai, M. (2010). The mechanism of rotating proton pumping ATPases. *Biochimica et Biophysica Acta (BBA)—Bioenergetics* 8, 1343–1352. https://doi.org/10.1016/j.bbabio.2010.02.014

Nakamoto, R. K., Scanlon, J. A. B., & Al-Shawi, M. K. (2008). The rotary mechanism of the ATP synthase. *Archives of Biochemistry and Biophysics* 476, 43–50. https://doi.org/10.1016/j.abb.2008.05.004

Nepi, M. (2014). Beyond nectar sweetness: the hidden ecological role of non-protein amino acids in nectar. *Journal of Ecology* 102, 108–115. https://doi.org/10.1111/1365-2745.12170

Nepi, M., Grasso, D. A., & Mancuso, S. (2018). Nectar in plant-insect mutualistic relationships: from food reward to partner manipulation. *Frontiers in Plant Science* 9, 1063. https://doi.org/10.3389/fpls.2018.01063

Ng, W.-K. (2017). Justifying the precautionary principle with expected net-welfare maximization. *Animal Sentience* 2017, 040. https://doi.org/10.51291/2377-7478.1238

Nick, P. (2021). Sensitive or sentient—a painful debate. *Protoplasma* 258, 235–238. https://doi.org/10.1007/s00709-021-01621-5

Nicolson, G. L. (2014). The fluid–mosaic model of membrane structure: still relevant to understanding the structure, function and dynamics of biological membranes after more than 40 years. *Biochimica et Biophysica Acta* 1838, 1451–1466. https://doi.org/10.1016/j.bbamem.2013.10.019

Nicolson, G. L. & Ferreira de Mattos, G. (2022). Fifty years of the fluid-mosaic model of biomembrane structure and organization and its importance in biomedicine with particular emphasis on membrane lipid replacement. *Biomedicines* 10, 1711. https://doi.org/10.3390/biomedicines10071711

Nieder, A., Wagener, L., & Rinnert, P. (2020). A neural correlate of sensory consciousness in a corvid bird. *Science* 369, 1626–1629. https://doi.org/10.1126/science.abb1447

Niklas, K. J. & Newman, S. A. (2020). The many roads to and from multicellularity. *Journal of Experimental Botany* 71, 3247–3253. OI:10.1093/jxb/erz547

Nirody, J. A., Budin, I., & Rangamani, P. (2020). ATP synthase: evolution, energetics, and membrane interactions. *Journal of General Physiology* 152, e201912475. https://doi.org/10.1085/jgp.201912475

Noble, D. (2017). Evolution viewed from physics, physiology and medicine. *Interface Focus* 7, 20160159. https://doi.org/10.1098/rsfs.2016.0159

Noble, D. (2021). The illusions of the modern synthesis. *Biosemiotics* 14(1), 5–24.

Noble, R. & Noble, D. (2018). Harnessing stochasticity: how do organisms make choices? *Chaos: An Interdisciplinary Journal of Nonlinear Science* 28, 106309. https://doi.org/10.1063/1.5039668

Nobs, S. J., MacLeod, F. I., Wong, H. L., & Burns, B. P. (2022). Eukarya the chimera: eukaryotes, a secondary innovation of the two domains of life? *Trends in Microbiology* 30, 421–431. https://doi.org/10.1016/j.tim.2021.11.003

Noctor, G. & Foyer, C. H. (2016). Intracellular redox compartmentation and ROS-related communication in regulation and signaling. *Plant Physiology* 171. 1581–1592. https://doi.org/10.1104/pp.16.00346

Nohr, D., Franz, S., Rodriguez, R., Paulus, B., Essen, L. O., Weber, S., et al. (2016). Extended electron-transfer in animal cryptochromes mediated by a tetrad of aromatic amino acids. *Biophysics Journal* 111, 301–311. https://doi.org/10.1016/j.bpj.2016.06.009

Norris, V. (2021). Competitive coherence generates *qualia* in bacteria and other living systems. *Biology* 10, 1034. https://doi.org/10.3390/biology10101034

Norstog, K. J., Gifford, E. M., & Stevenson, D. W. (2004). Comparative development of the spermatozoids of Cycads and *Gingko biloba*. *Botanical Reviews* 70, 5–15. https://doi.org/10.1663/0006-8101(2004)070[0005:CDOTSO]2.0.CO;2

Novoplansky, A. (2009). Picking battles wisely: plant behaviour under competition. *Plant & Cell Environment* 32, 726–741. https://doi.org/10.1111/j.1365-3040.2009.01979.x

Nowak, M. A. (2006). Five rules for the evolution of cooperation. *Science* 314, 1560–1563. https://doi.org/10.1126/science.1133755

Nuccitelli, R. & Jaffe, L. F. (1974). Spontaneous current pulses through developing fucoid eggs. *Proceedings of the National Academy of Sciences USA* 71, 4855–4859. https://doi.org/10.1073/pnas.71.12.4855

Nurse, P. (2020). *What Is Life?: Understand Biology in Five Steps*. Oxford: David Fickling Books.

Nutman, A. P., Bennett, V. C., Friend, C. R. L., Van Kranendonk, M. J., Rothacker, L., & Chivas, A. R. (2019). Cross-examining Earth's oldest stromatolites: seeing through the effects of heterogeneous deformation, metamorphism and metasomatism affecting Isua (Greenland) ≈ 3700 Ma sedimentary rocks. *Precambrian Research* 331, 105347. https://doi.org/10.1016/j.precamres.2019.105347

Oba, Y., Takano, Y., Furukawa, Y., Glavin, D. P., Dworkin, J. P., & Naraoka, H. (2022). Identifying the wide diversity of extraterrestrial purine and pyrimidine nucleobases in carbonaceous meteorites. *Nature Communications* 13, 2008. https://doi.org/10.1038/s41467-022-29612-x

Oberkofler, V., Pratx, L., & Bäurle, I. (2021). Epigenetic regulation of abiotic stress memory: maintaining the good things while they last. *Current Opinion in Plant Biology* 61, 102007. https://doi.org/10.1016/j.pbi.2021.102007

Ogas, O. & Gaddam, S. (2022). *Journey of the Mind: How Thinking Emerged From Chaos.* New York: Norton.

O'Hara-Wright, M., Mobini, S., & Gonzalez-Cordero, A. (2022). Bioelectric potential in next-generation organoids: electrical stimulation to enhance 3D structures of the central nervous system. *Frontiers in Cellular and Developmental Biology* 10, 901652. https://doi.org/10.3389/fcell.2022.901652

Oizumi, M., Albantakis, L., & Tononi, G. (2014). From the phenomenology to the mechanisms of consciousness: integrated information theory 3.0. *PLoS Computational Biology* 10, e1003588. https://doi.org/10.1371/journal.pcbi.1003588

Oliveira-Brett, A. M. (2017). Electron transfer reactions in biological systems. In: *Reference Module in Chemistry, Molecular Sciences and Chemical Engineering, pp.* 1–8. [Online] Elsevier. https://doi.org/10.1016/B978-0-12-409547-2.12790-8

Oliver, K. R. & Greene, W. K. (2011). Mobile DNA and the TE-thrust hypothesis: supporting evidence from the primates. *Mobile DNA* 2, 8. https://doi.org/10.1186/1759-8753-2-8

Okazaki, N., Takai, K., & Sato, T. (1993). Immobilization of a sensitive plant, *Mimosa pudica* L., by volatile anesthetics. *Masui* 42, 1190–1193. PMID: 8366560.

O'Malley, M. A., Leger, M. M., Wideman, J. G., & Ruiz-Trillo, I. (2019). Concepts of the last eukaryotic common ancestor. *Nature Ecology & Evolution* 3, 338–344. https://doi.org/10.1038/s41559-019-0796-3

Omar, F., Tareq, A. M., Alqahtani, A. M., Daman, K., Emran, T. B., & Simal-Gandara, J. (2021). Plant-based indole alkaloids: a comprehensive overview from a pharmacological perspective. *Molecules* 26, 2297. https://doi.org/10.3390/molecules26082297

O'Neil, S. D., Rácz, B., Brown, W. A., Gao, Y., Soderblom, E. J., Yasuda, R., et al. (2021). Action potential-coupled Rho GTPase signaling drives presynaptic plasticity. *eLife* 10, e63756. https://doi.org/10.7554/eLife.63756

O'Neill, J. S. (2021). Redox-coupled rhythm and brews. *Nature Chemical Biology* 17, 373–374. https://doi.org/10.1038/s41589-021-00777-8

Orioli, G., Santoni, A., Dragovic, D., & Farroni, T. (2019). Identifying peripersonal space boundaries in newborns. *Scientific Reports* 9, 9370. https://doi.org/10.1038/s41598-019-45084-4

Osterhout, W. J. V. (1952). Some aspects of protoplasmic motion. *Journal of General Physiology* 35, 519–527. https://doi.org/10.1085/jgp.35.3.519

Overgaard, M. (2021). Insect consciousness. *Frontiers in Behavioral Neuroscience* 15, 653041. https://doi.org/10.3389/fnbeh.2021.653041

Overton, E. (1901). *Studien über die Narkose.* Jena: Fischer.

Palanivelu, R. & Preuss, D. (2001). Pollen tube targeting and axon guidance: parallels in tip growth mechanisms. *Trends in Cell Biology* 10, 517–524. https://doi.org/10.1016/S0962-8924(00)01849-3

Palanivelu, R. & Tsukamoto, T. (2011). Pathfinding in angiosperm reproduction: pollen tube guidance by pistils ensures successful double fertilization. *Wiley Interdisciplinary Reviews Developmental Biology* 1, 96–113. https://doi.org/10.1002/wdev.6

Palanivelu, R., Brass, L., Edlund, A. F., & Preuss, D. (2003). Pollen tube growth and guidance is regulated by POP2, an *Arabidopsis* gene that controls GABA levels. *Cell* 114, 47–59. https://doi.org/10.1016/s0092-8674(03)00479-3

Palmer, J. D. & Foster, K. R. (2022). Bacterial species rarely work together. *Science* 376, 581–582.

Pan, J. A., Peng, X., Gao, Y., Li, Z., Lu, X., Chen, Y., et al. (2008). Genome-wide analysis of protein-protein interactions and involvement of viral proteins in SARS-CoV replication. *PLoS One* 3, e3299. https://doi.org/10.1186/s13578-021-00644-y

Papineau, D., She, Z., Dodd, M. S., Iacoviello, F., Slack, J. F., Hauri, E., et al. (2022). Metabolically diverse primordial microbial communities in Earth's oldest seafloor-hydrothermal jasper. *Science Advances* 8, 2296. https://doi.org/10.1126/sciadv.abm2296

Papir-Kricheli, D. & Magnes, J. (1982). Electroanesthesia in the cat: standardization of electric current parameters and anesthetic criteria. *Israel Journal of Medical Sciences* 18, 99–103. PMID: 7068348.

Parise, A. G., Reissig, G. N., Basso, L. F., Senko, L. G. S., Oliveira, T. F. C., de Toledo, G. R. A., et al. (2021. Detection of different hosts from a distance alters the behaviour and bioelectrical activity of *Cuscuta racemosa*. *Frontiers in Plant Science* 12, 594195. https://doi.org/10.3389/fpls.2021.594195

Parry, B. R., Surovtsev, I. V., Cabeen, M. T., O'Hem, C. S., Dufresne, E. R., & Jacobs-Wagner, C. (2013). The bacterial cytoplasm has glass-like properties and is fluidized by metabolic activity. *Cell* 156, 183–194. https://doi.org/10.1016/j.cell.2013.11.028

Patel, A., Malinovska, L., Saha, S., Wang, J., Alberti, S., Krishnan, Y., et al. (2017). ATP as a biological hydrotrope. *Science* 356, 753–756. https://doi.org/10.1126/science.aaf6846

Paulose, J. K., Wright, J. M., Patel, A. G., & Cassone, V. M. (2016). Human gut bacteria are sensitive to melatonin and express endogenous circadian rhythmicity. *PLoS One* 11, e0146643. https://doi.org/10.1371/journal.pone.0146643

Pavlovič, A., Libiaková, M., Bokor, B., Jakšová, J., Petřík, I., Novák, O., et al. (2020). Anaesthesia with diethyl ether impairs jasmonate signalling in the carnivorous plant Venus flytrap (*Dionaea muscipula*). *Annals of Botany* 125, 173–183. https://doi.org/10.1093/aob/mcz177

Peinkhofer, C., Martial, C., Cassol, H., Laureys, S., & Kondziella, D. (2021). The evolutionary origin of near-death experiences: a systematic investigation. *Brain Communications* 3, fcab132. https://doi.org/10.1093/braincomms/fcab132

Peiter, E. (2016). The ever-closer union of signals: propagating waves of calcium and ROS are inextricably linked. *Plant Physiology* 172, 3–4. https://doi.org/10.1104/pp.16.01037

Peng, X., Liu, Y., He, W., Hoppe, E. D., Zhou, L., Xin, F., Haswell, E. S., Pickard, B. G., Genin, G. M., & Lu T. J. (2022). Acoustic radiation force on a long cylinder, and potential sound transduction by tomato trichomes. *Biophysical Journal* 121, 3917–3926. https://doi.org/10.1016/j.bpj.2022.08.038

Pereira, C. (2015). Is it quantum sentience or quantum consciousness? A review of social behaviours observed in primitive and present-day microorganisms. *NeuroQuantology* 14, 16–27. https://doi.org/10.14704/nq.2016.14.1.874

Perouansky, M. (2012). The quest for a unified model of anesthetic action: a century in Claude Bernard's shadow. *Anesthesiology* 117, 465–474. https://doi.org/10.1097/ALN.0b013e318264492e

Perouansky, M. (2015). The Overton in Meyer-Overton: a biographical sketch commemorating the 150th anniversary of Charles Ernest Overton's birth. *British Journal of Anaesthesia* 114, 537–541. https://doi.org/10.1093/bja/aev069

Picciotto, S., Santonicola, P., Paterna, A., Rao, E., Raccosta, S., Romancino, D. P., et al. (2022). Extracellular vesicles from microalgae: uptake studies in human cells and *Caenorhabditis elegans*. *Frontiers in Bioengineering and Biotechnology* 10, 830189. https://doi.org/10.3389/fbioe.2022.830189

Pierce, S. K. & Curtis, N. E. (2012). Cell biology of the chloroplast symbiosis in sacoglossan sea slugs. *International Reviews of Cell Molecular Biology* 293, 123–148. https://doi.org/10.1016/B978-0-12-394304-0.00009-9

Pietak, A. & Levin, M. (2018). Bioelectrical control of positional information in development and regeneration: a review of conceptual and computational advances. *Progress in Biophysics and Molecular Biology* 137, 52–68. https://doi.org/10.1016/j.pbiomolbio.2018.03.00

Pigliucci, M. (2007). Do we need an extended evolutionary synthesis? *Evolution: International Journal of Organic Evolution* 61, 2743–2749. https://doi.org/10.1111/j.1558-5646.2007.00246.x

Pinello, J. F. & Clark, T. G. (2022). HAP2-mediated gamete fusion: lessons from the world of unicellular eukaryotes. *Frontiers of Cellular and Developmental Biology* 9, 807313. https://doi.org/10.3389/fcell.2021.807313

Pinello, J. F., Lai, A. L., Millet, J. K., Cassidy-Hanley, D., Freed, J. H., & Clark, T. G. (2017). Structure-function studies link class II viral fusogens with the ancestral gamete fusion protein HAP2. *Current Biology* 27, 651–660. https://doi.org/10.1016/j.cub.2017.01.049

Pirozynski, K. A. & Malloch, D. W. (1975). The origin of land plants: a matter of mycotrophism. *Biosystems* 6, 153–164. https://doi.org/10.1016/0303-2647(75)90023-4

Platholi, J., Herold, K. F., Hemmings, H. C., Jr., & Halpain, S. (2014). Isoflurane reversibly destabilizes hippocampal dendritic spines by an actin-dependent mechanism. *PLoS One* 9, e102978. https://doi.org/10.1371/journal.pone.0102978

Platre, M. P. & Jaillais, Y. (2017). Anionic lipids and the maintenance of membrane electrostatics in eukaryotes. *Plant Signaling & Behavior* 12, e1282022. https://doi.org/10.1080/15592324.2017.1282022

Platre, M. P., Noack, L. C., Doumane, M., Bayle, V., Simon, M. L. A., Maneta-Peyret, L., et al. (2018). A combinatorial lipid code shapes the electrostatic landscape of plant endomembranes. *Developmental Cell* 45, 465–480. https://doi.org/10.1016/j.devcel.2018.04.011

Plikynas, D., Masteika, S., & Budrionis, A. (2012). Interdisciplinary principles of field-like coordination in the case of self-organized social systems. *International Journal of Computer and Information Engineering* 61, 754–759.

Podbilewicz, B. (2014). Virus and cell fusion mechanisms. *Annual Reviews of Cell and Developmental Biology* 30, 111–139. https://doi.org/10.1146/annurev-cellbio-101512-122422

Poirot, O. & Timsit, Y. (2016). Neuron-like networks between ribosomal proteins within the ribosome. *Scientific Reports* 6, 26485. https://doi.org/10.1038/srep26485

Pollack, G. H. (2001). *Cells, Gels and the Engines of Life: A New, Unifying Approach to Cell Function*. Seattle, WA: Ebner and Sons Publishers.

Pollack, G. H. (2013). *The Fourth Phase of Water: Beyond Solid, Liquid, and Vapor*. Seattle, WA: Ebner and Sons Publishers.

Polo, S. & Di Fiore, P. P. (2006). Endocytosis conducts the cell signaling orchestra. *Cell* 124, 897–900. https://doi.org/10.1016/j.cell.2006.02.025

Ponte, G., Chiandetti, C., Edelman, D. B., Imperadore, P., Pieroni, E. M., & Fiorito, G. (2022). Cephalopod behavior: from neural plasticity to consciousness. *Frontiers in Behavioral Neuroscience* 15, 787139. https://doi.org/10.3389/fnsys.2021.787139

Porter, K. R. (1984). The cytomatrix: a short history of its study. *Journal of Cell Biology* 99, 3–12. https://doi.org/10.1083/jcb.99.1.3s

Porter, S. M. (2020). Insights into eukaryogenesis from the fossil record. *Interface Focus* 10, 20190105. https://doi.org/10.1098/rsfs.2019.0105

Portes, A. T. & Feijó, J. A. (2021). Measuring extracellular proton and anionic fluxes in Arabidopsis pollen tubes. *Bio-Protocol* 11, e3908. https://doi.org/10.21769/BioProtoc.3908

Portmore, D. W. (2011). *Commonsense Consequentialism: Wherein Morality Meets Rationality*. New York: Oxford University Press.

Pospíšil, P., Kumar, A., & Prasad, A. (2022). Reactive oxygen species in photosystem II: relevance for oxidative signaling. *Photosynthesis Research* 152, 245–260. https://doi.org/10.1007/s11120-022-00922-x

Postler, T. S. & Ghosh, S. (2017). Understanding the holobiont: how microbial metabolites affect human health and shape the immune system. *Cell Metabolism* 26, 110–130. https://doi.org/10.1016/j.cmet.2017.05.008

Prado, A. M., Colaço, R., Moreno, N., Silva, A. C., & Feijó, J. A. (2011). Targeting of pollen tubes to ovules is dependent on nitric oxide (NO) signaling. *Molecular Plant* 1, 703–714. https://doi.org/10.1093/mp/ssn034

Prebble, J. N. (2001). The philosophical origins of Mitchell's chemiosmotic concepts: the personal factor in scientific theory formulation. *Journal of the History of Biology* 34, 433–460. https://doi.org/10.1023/a:1012946715673

Prigogine, I. & Stengers, I. (1984). *Order Out of Chaos*. London: Bantam Books Inc.

Prindle, A., Liu, J., Asally, M., Ly, S., Garcia-Ojalvo, J., & Süel, G. M. (2015). Ion channels enable electrical communication in bacterial communities. *Nature* 527, 59–63. https://doi.org/10.1038/nature15709

Prinz, W. A., Toulmay, A., & Balla, T. (2020). The functional universe of membrane contact sites. *Nature Reviews Molecular & Celular Biology* 21, 7–24. https://doi.org/10.1038/s41580-019-0180-9

Pross, A. (2016). *What is Life?: How Chemistry Becomes Biology*, 2nd ed. Oxford: Oxford University Press.

Putnam, H. (1999). *The Threefold Cord: Mind, Body, and World*. New York: Columbia University Press.

Qiao, Y., Li, M., Booth, R., & Mann, S. (2017). Predatory behaviour in synthetic protocell communities. *Nature Chemistry* 9, 110–119. https://doi.org/10.1038/nchem.2617

Qiu, G. H., Yang, X., Zheng, X., & Huang, C. (2017). The eukaryotic genome is structurally and functionally more like a social insect colony than a book. *Epigenomics* 9, 1469–1483. https://doi.org/10.2217/epi-2017-0059

Raanan, H., Pike, D. H., Moore, E. K., Falkowski, P. G., & Nanda, V. (2018). Modular origins of biological electron transfer chains. *Proceedings of the National Academy of Sciences USA* 115, 1280–1285. https://doi.org/10.1073/pnas.1714225115

Raanan, H., Poudel, S., Pike, D. H., Nanda, V., & Falkowski, P. G. (2020). Small protein folds at the root of an ancient metabolic network. *Proceedings of the National Academy of Sciences USA* 117, 7193–7199. https://doi.org/10.1073/pnas.1914982117

Rabellino, D., Frewen, P. A., McKinnon, M. C., & Lanius, R. A. (2020). Peripersonal space and bodily self-consciousness: implications for psychological trauma-related disorders. *Frontiers in Neuroscience* 14, 586605. https://doi.org/10.3389/fnins.2020.586605

Rabouille, C., Cortassa, S., & Aon, M. A. (1992). Fractal organisation in bio-logical macromolecular lattices. *Journal of Biomolecular Structure and Dynamics* 9, 1013–1024. https://doi.org/10.1080/07391102.1992.10507973

Rácz, B., Dušková, M., Stárka, L., Hainer, V., & Kunešová, M. (2018). Links between the circadian rhythm, obesity and the microbiome. *Physiological Research* 67, S409–S420.

Rad, I., Stahlberg, R., Kung, K., & Pollack, G. H. (2021). Low frequency weak electric fields can induce structural changes in water. *PLoS One* 16, e0260967. https://doi.org/10.1371/journal.pone.0260967

Ram, M. & Babbar, S. B. (2002). Transient existence of life without a cell membrane: a novel strategy of siphonous seaweed for survival and propagation. *BioEssays* 24, 588–590. https://doi.org/10.1002/bies.10117

Ramesh, M. A., Malik, S. B., & Logsdon, J. M., Jr. (2005). A phylogenomic inventory of meiotic genes; evidence for sex in Giardia and an early eukaryotic origin of meiosis. *Current Biology* 15, 185–191. https://doi.org/10.1016/j.cub.2005.01.003

Ramón y Cajal, S. (1894). La fine structure des centres nerveux. The Croonian lecture. *Proceedings of the Royal Society B: Biological Sciences*. 55, 443–468.

Ratsliff, W. C., Herron, M. D., Howell, K., Pentz, J. T., Rosenzweig, F., & Travisano, M. (2013). Experimental evolution of an alternating uni- and multicellular life cycle in *Chlamydomonas reinhardtii*. *Nature Communication* 4, 2742. https://doi.org/10.1038/ncomms3742

Reber, A. S. (2011). An epitaph for grammar. In: Sanz, C. & Loew, R. P. (Eds.), *Implicit and Explicit Language Learning*, pp. 23–34. Washington, DC: Georgetown University Press.

Reber, A. S. (2017). What if all animals are sentient? A commentary on Birch on the Precautionary Principle. *Animal Sentience* 16, 029.

Reber, A. S. (2019). *The First Minds: Caterpillars, Karyotes, and Consciousness*. New York: Oxford University Press.

Reber, A. S. & Allen, R. (Eds.) (2022). *The Cognitive Unconscious: The First Half-Century*. New York: Oxford University Press.

Reber, A. S. & Baluška, F. (2021). Cognition in some surprising places. *Biochemical and Biophysical Research Communications* 564, 150–157. https://doi.org/10.1016/j.bbrc.2020.08.115

Reber, A. S. & Baluška, F. (2022). Where minds begin: a commentary on Joseph LeDoux's *The Deep History of Ourselves*. *Philosophical Psychology*. Published online 28 December 2022. https://doi.org/10.1080/09515089.2022.2160700

Reber, A. S., Baluška, F., & Miller, W. B., Jr. (2022). Of course crustaceans are sentient, but there's more to the story: comment on Crump, et al. *Animal Sentience* 32, 3. https://doi.org/10.51291/2377-7478.1712

Reber, A. S., Baluška, F., & Miller, W. B., Jr. (In press). The sentient cell. In: Sharov, A., Mikhailovsky, G., Gordon, R., & Seckbach, J. (Eds.), *Pathways to the Origin and Evolution of Meanings in the Universe*. Beverly, MA: Scrivener.

Reddy, J. S. K. (2017). Understanding the emergence of microbial consciousness: from a perspective of the subject-object model. *Journal of Integrative Neuroscience* 16, 27–36. https://doi.org/10.3233/JIN-170064

Reddy, J. S. K. & Pereira, C. (2016). Origin of life: a consequence of cosmic energy, redox homeostasis and quantum phenomenon. *NeuroQuantology* 14, 581–588. https://doi.org/10.14704/nq.2016.14.3.914

Reiter, R. J., Mayo, J. C., Tan, D. X., Sainz, R. M., Moises Alatorre-Jimenez, M., & Qin, L. (2016). Melatonin as an antioxidant: under promises but over delivers. *Journal of Pineal Research* 61, 253–278. https://doi.org/10.1111/jpi.12360

Reiter, R. J., Tan, D., Rosales.Corrla, S., Galano. A., Zhou, X. J., & Xu, B. (2018). Mitochondria: central organelles for melatonin's antioxidant and anti-aging actions. *Molecules* 23, 509. https://doi.org/10.3390/molecules23020509

Remy, W., Taylor, T. N., Hass, H., & Kerp, H. (1994). Four hundred-million-year-old vesicular-arbuscular mycorrhizae. *Proceedings of the National Academy of Sciences USA* 91, 11841–11843. https://doi.org/10.1073/pnas.91.25.11841

Renzaglia, K. S. & Garbary, D. J. (2001). Motile gametes of land plants: diversity, development, and evolution. *Critical Reviews of Plant Sciences* 20, 107–213. https://doi.org/10.1080/20013591099209

Revah, O., Gore, F., Kelley, K. W., Andersen, J., Sakai, N., Chen, X., et al. (2022). Maturation and circuit integration of transplanted human cortical organoids. *Nature* 610, 319–326. https://doi.org/10.1038/s41586-022-05277-w

Rey, G. & Reddy, A. B. (2015). Interplay between cellular redox oscillations and circadian clocks. *Diabetes Obesity and Metabolism* 17 Suppl. 1, 55–64. https://doi.org/10.1111/dom.12519

Reynaud, K., Schuss, Z., Rouach, N., & Holcman, D. (2015). Why so many sperm cells? *Communicative & Integrative Biology* 8, e1017156. https://doi.org/10.1080/19420889.2015.1017156

Reynierse, J. H. & Walsh, G. L. (1967). Behavior modification in the protozoan Stentor re-examined. *Psychological Record* 17, 161–165. https://link.springer.com/article/10.1007/BF03393700

Rice, A. M. & Rosen, M. K. (2017). ATP controls the crowd. *Science* 356, 701–702. https://doi.org/10.1126/science.aan4223

Richardson, K. H., Wright, J. J., Šimėnas, M., Thiemann, J., Esteves, A. M., McGuire, G., et al. (2021). Functional basis of electron transport within photosynthetic complex I. *Nature Communications* 12, 5387. https://doi.org/10.1038/s41467-021-25527-1

Rinaldi, A. (2014). Reawakening anaesthesia research. *EMBO Reports* 15, 1113–1118. https://doi.org/10.15252/embr.201439593

Riol, A., Cervera, J., Levin, M., & Mafe, S. (2021). Cell systems bioelectricity: how different intercellular gap junctions could regionalize a multicellular aggregate. *Cancers (Basel)* 13, 5300. https://doi.org/10.3390/cancers13215300

Rivas, G., Ferrone, F., & Herzfeld, J. (2004). Life in a crowded world. *EMBO Reports* 5, 23–27. https://doi.org/10.1038/sj.embor.7400056

Rivera-Milla, E., Stuermer, C. A., & Málaga-Trillo, E. (2006). Ancient origin of reggie (flotillin), reggie-like, and other lipid-raft proteins: convergent evolution of the SPFH domain. *Cellular and Molecular Life Sciences* 63, 343–357. https://doi.org/10.1007/s00018-005-54343

Riveros-Pereza, E. & Riveros, R. (2018). Water in the human body: an anesthesiologist's perspective on the connection between physicochemical properties of water and physiologic relevance. *Annals of Medicine and Surgery* 26, 1–8. https://doi.org/10.1016/j.amsu.2017.12.007

Rodrigo-Moreno, A., Bazihizina, N., Azzarello, E., Masi, E., Tran, D., Bouteau, F., et al. (2017). Root phonotropism: early signalling events following sound perception in Arabidopsis roots. *Plant Science* 264, 9–15. https://doi.org/10.1016/j.plantsci.2017.08.001

Rodrigues-Oliveira, T., Wollweber, F., Ponce-Toledo, R. I., Xu, J., Rittmann, S. K. M., Klingl, A., Pilhofer, M., & Schleper, C. (2023). Actin cytoskeleton and complex cell architecture in an Asgard archaeon. *Nature* 613, 332–339. https://doi.org/10.1038/s41586-022-05550-y

Roederer, J. G. (2004). When and where did information first appear in the Universe? In: Seckbach, J. & Rub, E. (Eds.), *The New Avenues in Bioinformatics*, pp. 23–37. Dordrecht: Springer.

Rogers, S. M. & Simpson, S. J. (2014). Thanatosis. *Current Biology* 24, R1031–R1033. https://doi.org/10.1016/j.cub.2014.08.051

Ross, W. N. (2012). Understanding calcium waves and sparks in central neurons. *Nature Reviews Neuroscience* 13, 157–168. https://doi.org/10.1038/nrn3168

Rosset, S. L., Oakley, C. A., Ferrier-Pagès, C., Suggett, D. J., Weis, V. M., & Davy, S. K. (2021). The molecular language of the cnidarian-dinoflagellate symbiosis. *Trends in Microbiology* 29, 320–333. https://doi.org/10.1016/j.tim.2020.08.005

Rossini, M., Pizzo, P., & Filadi, R. (2021). Better to keep in touch: investigating inter-organelle cross-talk. *FEBS Journal* 288, 740–755. https://doi.org/10.1111/febs.15451

Rozsa, M. (2022). Can plants think? The burgeoning field of plant neurobiology has a lot to say on the matter. *Salon* September. https://www.salon.com/2022/09/30/can-plants-think-the-burgeoning-field-of-plant-neurobiology-has-a-lot-to-say-on-the-matter/

Rumpho, M. E., Summer, E. J., & Manhart, J. R. (2000). Solar-powered sea slugs. Mollusc/algal chloroplast symbiosis. *Plant Physiology* 123, 29–38. https://doi.org/10.1104/pp.123.1.29

Rumpho, M. E., Pelletreau, K. N., Moustafa, A., & Bhattacharya, D. (2011). The making of a photosynthetic animal. *Journal of Experimental Biology* 214, 303–311. https://doi.org/10.1242/jeb.046540

Runyon, J. B., Mescher, M. C., & De Moraes, C. M. (2006). Volatile chemical cues guide host location and host selection by parasitic plants. *Science* 313, 1964–1967. https://doi.org/10.1126/science.1131371

Sagan, L. (1967). On the origin of mitosing cells. *Journal of Theoretical Biology* 14, 255–274. https://doi.org/10.1016/0022-5193(67)90079-3

Saita, E. I., Suzuki, T., Kinosita, K., Jr., & Yoshida, M. (2015). Simple mechanism whereby the F1-ATPase motor rotates with near-perfect chemomechanical energy conversion. *Proceedings of the National Academy of Sciences USA* 112, 9626–9631. https://doi.org/10.1073/pnas.1422885112

Salomon, R., Noel, J. P., Łukowska, M., Faivre, N., Metzingerm, T., Serino, A., et al. (2017). Unconscious integration of multisensory bodily inputs in the peripersonal space shapes bodily self-consciousness. *Cognition* 166, 174–183. https://doi.org/10.1016/j.cognition.2017.05.028

Šamaj, J., Read, N. D., Volkmann, D., Menzel, D., & Baluška, F. (2005). The endocytic network in plants. *Trends in Cell Biology* 15, 425–433. https://doi.org/10.1016/j.tcb.2005.06.006

Sandberg, W. S. & Miller, K. W. (2003). The Meyer-Overton relationship and its exceptions. In: Antognini, J. F. (Ed.), *Contemporary Clinical Neuroscience: Neural Mechanisms of Anesthesia*, pp. 371–394. Totowa, NJ: Humana Press Inc.

Sanjuán, R. & Domingo-Calap, P. (2016). Mechanisms of viral mutation. *Cellular and Molecular Life Science* 73, 4433–4448. https://doi.org/10.1007/s00018-016-2299-6

Santolini, J., Wootton, S. A., Jackson, A. A., & Feelisch, M. (2019). The redox architecture of physiological function. *Current Opinion in Physiology* 9, 34–47. https://doi.org/10.1016/j.cophys.2019.04.00

Sarkar, S., Das, S., Dagar, S., Joshi, M. P., Mungi, C. V., Sawant, A. A., et al. (2020). Prebiological membranes and their role in the emergence of early cellular life. *Journal of Membrane Biology* 253, 589–608. https://doi.org/10.1007/s00232-020-00155-w

Sarkar, S. & Mondal, J. (2021). Mechanistic insights on ATP's role as a hydrotrope. *Journal of Physical Chemistry B* 125, 7717–7731. https://doi.org/10.1021/acs.jpcb.1c03964

Sathe, S. & Durand, P. M. (2016). Cellular aggregation in *Chlamydomonas* (Chlorophyceae) is chimaeric and depends on traits like cell size and motility. *European Journal of Phycology* 51, 129–138. https://doi.org/10.1080/09670262.2015.1107759

Savtchenko, L. P., Poo, M. M., & Rusakov, D. A. (2017). Electrodiffusion phenomena in neuroscience: a neglected companion. *Nature Reviews Neurosciences* 18, 598–612. https://doi.org/10.1038/nrn.2017.101

Sazanov, L. A. (2015). A giant molecular proton pump: structure and mechanism of respiratory complex I. *Nature Reviews Molecular and Cell Biology* 16, 375–388. https://doi.org/10.1038/nrm3997

Schaap, P. (2021). From environmental sensing to developmental control: cognitive evolution in dictyostelid social amoebas. *Philosophical Transaction of the Royal Society London B Biological Sciences* 376, 20190756. https://doi.org/10.1098/rstb.2019.0756

Schatten, H., Rawe, V. Y., & Sun, Q. Y. (2011). The sperm centrosome: its role and significance in nature and human assisted reproduction. *Journal of Reproductive Biotechnology and Fertility* 2, 121–127. https://doi.org/10.1177/205891581100200206

Schertzer, J. W., Bouletter, M. L., & Whiteley, M. (2009). More than a signal: n-signaling properties of quorum sensing molecules. *Trends in Microbiology* 17, 189–195. https://doi.org/10.1016/j.tim.2009.02.001

Scherzer, S., Huang, S., Iosip, A., Kreuzer, I., Yokawa, K., Al-Rasheid, K. A. S., et al. (2022). Ether anesthetics prevents touch-induced trigger hair calcium-electrical signals excite the Venus flytrap. *Scientific Reports* 12, 2851. https://doi.org/10.1038/s41598-022-06915-z

Schiestl, F. P. (2005). On the success of a swindle: pollination by deception in orchids. *Naturwissenschaften* 92, 255–264. https://doi.org/10.1007/s00114-005-0636-y

Schmitt, D. L. & An, S. (2017). Spatial organization of metabolic enzyme complexes in cells. *Biochemistry* 56, 3184–3196. https://doi.org/10.1021/acs.biochem.7b00249

Schoemaecker, C. & Carslaw, N. (2022). The 'atmosphere' of the human body. *Science* 377, 1045–1046. https://doi.org/10.1126/science.add8461

Schrödinger, E. (1944). *What Is Life? The Physical Aspect of the Living Cell*. Cambridge: Cambridge University Press.

Schroeder, M. J. (2017). The difference that makes a difference for the conceptualization of information. *Multidisciplinary Digital Publishing Institute Proceedings* 1, 221.

Schumann, A. & Pancerz, K. (2016). *Physasum Chip: Growing Computers from Slime Mould. Logical aspects*. Szczecin, Poland: Scientific Publishing House IVG.

Schweinitzer, T. & Josenhans, C. (2010). Bacterial energy taxis: a global strategy? *Archives of Microbiology* 192, 507–520. https://doi.org/10.1007/s00203-010-0575-7

Schank, J. C. & Wimsatt, W. C. (1988). Generative entrenchment and evolution. In: Fine, A. & Machamer, P. (Eds.), *PSA 1986: Proceedings of the Meetings of the Philosophy of Science Association*, pp. 308–331. East Lansing, MI: Philosophy of Science Association.

Schumann, A. & Pancerz, K. (2016). *Physasum Chip: Growing Computers from Slime Mould. Logical Aspects*. Szczecin, Poland: Scientific Publishing House IVG.

Sciscioli, M., Lepore, E., Gherardi, M., & Scalera Liaci, L. (1994). Transfer of symbiotic bacteria in the mature oocyte of *Geodia cydonium* (Porifera, Demospongiae): an ultrastructural study. *Cahiers de Biologie Marine* 35, 471–478.

Scita, G. & Di Fiore, P. P. (2010). The endocytic matrix. *Nature* 463, 464–473. https://doi.org/10.1038/nature08910

Scott, B. I. H. (1967). Electric fields in plants. *Annual Review of Plant Physiology* 18, 409–418. https://doi.org/10.1146/annurev.pp.18.060167.002205

Scott, H. L., Bolmatov, D., Podar, P. T., Liu, Z., Kinnun, J. J., Doughty, B., et al. (2022). Evidence for long-term potentiation in phospholipid membranes. *Proceedings of the National Academy of Sciences USA* 119, e2212195119. https://doi.org/10.1073/pnas.2212195119

Searle, J. (1980). Minds, brains, and programs. *Behavioral and Brain Sciences* 3, 417–457.

Sekulić, D. L., Satarić, B. M., Tuszynski, J. A., & Satarić, M. V. (2011). Nonlinear ionic pulses along microtubules. *European Physical Journal E* 34, 49. https://doi.org/10.1140/epje/i2011-11049-0

Selosse, M. A. & Le Tacon, F. (1998). The land flora: a phototroph-fungus partnership? *Trends in Ecology & Evolution* 13, 15–20. https://doi.org/10.1016/s0169-5347(97)01230-5

Selye, H. (1975). *The Stress of Life*. New York: McGraw-Hill.

Sen, R. (2001). Budgeting for the wood-wide web. *New Phytologist* 145, 161–165. https://doi.org/10.1046/j.1469-8137.2000.00585.x

Sender, R., Fuchs, S., & Milo, R. (2016). Revised estimates for the number of human and bacteria cells in the body. *PLoS Biology* 14, e1002533. https://doi.org/10.1371/journal.pbio.1002533

Serino, A. (2019). Peripersonal space (PPS) as a multisensory interface between the individual and the environment, defining the space of the self. *Neuroscience & Biobehavioral Reviews* 99, 138–159. https://doi.org/10.1016/j.neubiorev.2019.01.016

Seyfarth, E. A. (2006). Julius Bernstein (1839–1917): pioneer neurobiologist and biophysicist. *Biological Cybernetics* 94, 2–8. https://doi.org/10.1007/s00422-005-0031-y

Shalatonin, V. & Pollack, G. H. (2022). Magnetic fields induce exclusion zones in water. *PLoS One* 17, e0268747. https://doi.org/10.1371/journal.pone.0268747

Shannon, C. (1948). The mathematical theory of communication. *Bell System Technical Journal* 27, 379–423.

Shapiro, J. A. (1992). Natural genetic engineering in evolution. *Genetica* 86, 99–111. https://doi.org/10.1007/BF00133714

Shapiro, J. A. (2007). Bacteria are small but not stupid: cognition, natural genetic engineering and socio-bacteriology. *Studies in the History, Philosophy of Biomedical Sciences* 38, 807–819. https://doi.org/10.1016/j.shpsc.2007.09.010

Shapiro, J. A. (2009). Revisiting the central dogma in the 21st century. *Annals of the New York Academy of Sciences* 1178, 6–28.

Shapiro, J. A. (2013). How life changes itself: the read–write (RW) genome. *Physics of Life Reviews* 10, 287–323. https://doi.org/10.1016/j.plrev.2013.07.001

Shapiro, J. A. (2016). The basic concept of the read–write genome: mini-review on cell-mediated DNA modification. *Biosystems* 140, 35–37. https://doi.org/10.1016/j.biosystems.2015.11.003

Shapiro, J. A. (2017). Biological action in read–write genome evolution. *Interface Focus* 7, 20160115.

Shapiro, J. A. (2019). No genome is an island: toward a 21st century agenda for evolution. *Annals of the New York Academy of Sciences* 1447, 21–52.

Shapiro, J. A. (2021). All living cells are cognitive. *Biochemical and Biophysical Research Communications* 564, 134–149. https://doi.org/10.1016/j.bbrc.2020.08.120

Sharp, C. & Foster, K. R. (2022). Host control and the evolution of cooperation in host microbiomes. *Nature Communications* 13, 3567. https://doi.org/10.1038/s41467-022-30971-8

Sharp, K. A. (2015). Analysis of the size dependence of macromolecular crowding shows that smaller is better. *Proceedings of the National Academy of Sciences USA* 112, 7990–7995. https://doi.org/10.1073/pnas.1505396112

Sheldon, R. B. (2005). Historical development of the distinction between bio- and abiogenesis. In: Hoover, R. B., Levin, G, V., Rozanov, A. Y., & Gladstone, G. R. (Eds.), *Astrobiology and Planetary Missions*, ol. 5906, p. 59061I. Ellingham, WA: Society of Photo-Optical Instrumentation Engineers.

Shemesh, H., Ovadia, O., & Novoplansky, A. (2010). Anticipating future conditions via trajectory sensitivity. *Plant Signaling & Behavior* 5, 1501–1503. https://doi.org/10.1371/journal.pone.0010824

Shepherd, V. A. (2006). The cytomatrix as a cooperative system of macromolecular and water networks. *Current Topics in Developmental Biology* 75, 171–223. https://doi.org/10.1016/S0070-2153(06)75006-2

Siddique, H. (2022). Give legal rights to animals, trees, and rivers, say experts. *The Guardian* 10 October. https://www.theguardian.com/environment/2022/oct/10/give-legal-rights-to-animals-trees-and-rivers-say-experts

Sigismund, S. & Scita, G. (2018). The endocytic matrix reloaded and its impact on the plasticity of migratory strategies. *Current Opinions in Cell Biology* 54, 9–17. https://doi.org/10.1016/j.ceb.2018.02.006

Sigismund, S., Confalonieri, S., Ciliberto, A., Polo, S., Scita, G., & Di Fiore, P. P. (2012). Endocytosis and signaling: cell logistics shape the eukaryotic cell plan. *Physiological Reviews* 92, 273–366. https://doi.org/10.1152/physrev.00005.2011

Simard, S. W. (2018). Mycorrhizal networks facilitate tree communication, learning and memory. In: Baluška, F., Gagliano, M., & Witzany, G. (Eds.), *Memory and Learning in Plants*, pp. 191–213. Cham: Springer Nature Switzerland AG.

Simon, M. L., Platre, M. P., Marquès-Bueno, M. M., Armengot, L., Stanislas, T., Bayle, V., et al. (2016). A PtdIns(4)P-driven electrostatic field controls cell membrane identity and signalling in plants. *Nature Plants* 2, 16089. https://doi.org/10.1038/nplants.2016.89

Simons, K. & Sampaio, J. L. (2011). Membrane organization and lipid rafts. *Cold Spring Harbor Perspectives in Biology* 3, a004697. https://doi.org/10.1101/cshperspect.a004697

Singer, S. J. & Nicolson, G. L. (1972). The fluid mosaic model of the structure of cell membranes. *Science* 175, 720–731. https://doi.org/10.1126/science.175.4023.720

Sjulstok, E., Olsen, J. M. H., & Solov'yov, I. A. (2015). Quantifying electron transfer reactions in biological systems: what interactions play the major role? *Scientific Reports* 5, 18446. https://doi.org/10.1038/srep18446

Skelhorn, J. (2018). Avoiding death by feigning death. *Current Biology* 28, R1135–R1136. https://doi.org/10.1016/j.cub.2018.07.070

Smith, E. & Morowitz, H. J. (2016). *The Origin and Nature of Life on Earth: The Emergence of the Fourth Geosphere*. Cambridge: Cambridge University Press. https://doi.org/10.1017/CBO9781316348772

Smith, J., Zadeh Haghighi, H., Salahub, D., & Simon, C. (2021). Radical pairs may play a role in xenon-induced general anesthesia. *Scientific Reports* 11, 6287. https://doi.org/10.1038/s41598-021-85673-w

Smith, S., Cianci, C., & Grima, R. (2017). Macromolecular crowding directs the motion of small molecules inside cells. *Journal of Royal Socienty—Interface* 14, 20170047. https://doi.org/10.1098/rsif.2017.0047

Snead, W. T. & Gladfelter, A. S. (2022). The control centers of biomolecular phase separation: now membrane surfaces, PTMs, and áctive processes regulate condensation. *Molecular Cell* 76, 295–305. https://doi.org/10.1016/j.molcel.2019.09.016

Sojo, V., Pomiankowski, A., & Lane, N. (2014). A bioenergetic basis for membrane divergence in archaea and bacteria. *PLoS Biology* 12, e1001926. https://doi.org/10.1371/journal.pbio.1001926

Song, Y. Y., Zeng, R. S., Xu, J. F., Li, J., Shen, X., & Yihdego, W. G. (2010). Interplant communication of tomato plants through underground common mycorrhizal networks. *PLoS One* 5, e13324. https://doi.org/10.1371/journal.pone.0013324

Song, Y. Y., Ye, M., Li, C., He, X., Zhu-Salzman, K., Wang, R. L., et al. (2014). Hijacking common mycorrhizal networks for herbivore-induced defence signal transfer between tomato plants. *Scientific Reports* 4, 3915. https://doi.org/10.1038/srep03915

Sonner, J. M. (2008). A hypothesis on the origin and evolution of the response to inhaled anesthetics. *Anesthesia & Analgesia* 107, 849–854. https://doi.org/10.1213/ane.0b013e31817ee684

Sonner, J. M. & Cantor, R. S. (2013). Molecular mechanisms of drug action: an emerging view. *Annual Reviews of Biophysics* 42, 143–167. https://doi.org/10.1146/annurev-biophys-083012-130341

Sorice, M., Misasi, R., Riitano, G., Manganelli, V., Martellucci, S., Longo, A., et al. (2021). Targeting lipid rafts as a strategy against coronavirus. *Frontiers in Cell and Developmental Biology* 8, 618296. https://doi.org/10.3389/fcell.2020.61829

Soucy, S. M., Huang, J., & Gogarten, J. P. (2015). Horizontal gene transfer: building the web of life. *Nature Review Genetics* 16, 472–482. https://doi.org/10.1038/nrg3962

Southworth, D. & Cresti, M. (1997). Comparison of flagellated and non-flagellated sperms in plants. *American Journal of Botany* 84, 1301–1311. https://www.jstor.org/stable/2446056

Spitzer, J. (2017). Emergence of life on Earth: a physicochemical jigsaw puzzle. *Journal of Molecular Evolution* 84, 1–7. https://doi.org/10.1007/s00239-016-9775-3

Srere, P. A. (2000). Macromolecular interactions: tracing the roots. *Trends in Biochemical Sciences* 25, 150–153. https://doi.org/10.1016/s0968-0004(00)01550-4

Staddon, M. F., Munro, E. M., & Banerjee, S. (2022). Pulsatile contractions and pattern formation in excitable actomyosin cortex. *PLoS Computational Biology* 18, e1009981. https://doi.org/10.1371/journal.pcbi.1009981

Stadhouders, R., Filion, G. J., & Graf, T. (2019). Transcription factors and 3D genome conformation in cell-fate decisions. *Nature* 569, 345–354. https://doi.org/10.1038/s41586-019-1182-7

Staehelin, A. L. (1997). The plant ER: a dynamic organelle composed of a large number of discrete functional domains. *Plant Journal* 11, 1151–1165. https://doi.org/10.1046/j.,1365-313x.1997.11061151.x

Stangherlin, A. & Reddy, A. B. (2013). Regulation of circadian clocks by redox homeostasis. *Journal of Biological Chemistry* 37, 26505–26511. https://doi.org/10.1074/jbc.R113.457564

Steinhardt, R. A., Bi, G., & Alderton, J. M. (1994). Cell membrane resealing by a vesicular mechanism similar to neurotransmitter release. *Science* 263, 390–394. https://doi.org/10.1126/science.7904084

Stewart, A. G., Sobti, M., Harvey, R. P., & Stock, D. (2013). Rotary ATPases: models, machine elements and technical specifications. *BioArchitecture* 3, 2–12. https://doi.org/10.4161/bioa.23301

Stewart, J. E. (2019). The origins of life: the managed-metabolism hypothesis. *Foundations of Science* 24, 171–195. https://doi.org/10.1007/s10699-018-9563-1

Stotz, H. U., Brotherton, D., & Inal, J. (2022). Communication is key: extracellular vesicles as mediators of infection and defence during host-microbe interactions in animals and plants. *FEMS Microbiology Reviews* 46, fuab044. https://doi.org/10.1093/femsre/fuab044

Strasburger, E. (1891). *Das Protoplasma und die Reizbarkeit. Rede zum Antritt des Rektorates der Rheinische Friedrich-Wilhelm Univesität Bonn.* Jena: von Gustav Fischer

Strauss, B. S. (2017). A physicist's quest in biology: Max Delbrück and 'complementarity'. *Genetics* 206, 641–650. https://doi.org/10.1534/genetics.117.201517

Striedter, G. (2016). Lack of neocortex does not imply fish cannot feel pain. *Animal Sentience* 3, 15.

Strippoli, P., Canaider, S., Noferini, F., D'Addabbo, P., Vitale, L., Facchin, F., et al. (2005). Uncertainty principle of genetic information in a living cell. *Theoretical Biology and Medical Modelling* 2, 1–6. https://doi.org/10.1186/1742-4682-2-40

Szabó, A. & Mayor, R. (2018). Mechanisms of neural crest migration. *Annual Review of Genetics* 52 43–63. https://doi.org/10.1146/annurev-genet-120417-031559

Sze, H. & Chanroj, S. (2018). Plant endomembrane dynamics: studies of K+/H+ antiporters provide insights on the effects of pH and ion homeostasis. *Plant Physiology* 117, 875–895. https://doi.org/10.1104/pp.18.00142

Szent-György, A. (1972). *The Living State.* New York: Academic Press.

Takagi, Y. A., Nguyen, D. H., Wexler, T. B., & Goldman, A. D. (2020). The coevolution of cellularity and metabolism following the origin of life. *Journal of Molecular Evolution* 88, 598–617. https://doi.org/10.1007/s00239-020-09961-1

Tagkopoulos, I., Liu, Y. C., & Tavazoie, S. (2008). Predictive behavior within microbial genetic networks. *Science* 320, 1313–1317. https://doi.org/10.1126/science.1154456

Tang, S. (2021). The origin(s) of cell(s): Pre-Darwinian evolution from FUCAs to LUCA. *Journal of Molecular Evolution* 89, 427–447. https://doi.org/10.1007/s00239-021-10014-4

Tauber, A. I. (2003). Metchnikoff and the phagocytosis theory. *Nature Reviews Molecular and Cellular Biology* 4, 897–901. https://doi.org/10.1038/nrm1244

Tauber, A. I. (2013). Immunology's theories of cognition. *History and Philosophy of the Life Sciences* 35, 239–264. PMID: 24466634.

Tauber, A. I. & Chernyak, L. (1991). *Metchnikoff and the Origins of Immunology: From Metaphor to Theory.* New York: Oxford University Press.

Thomas, L. (1978). *Lives of a Cell: Notes of a Biology Watcher.* New York: Penguin Books.

Thomas, M. A. & Cooper, R. L. (2022). Building bridges: mycelium-mediated plant-plant electrophysiological communication. *Plant Signaling & Behavior* 17, 2129291. https://doi.org/10.1080/15592324.2022.2129291

Thursby, E. & Juge, N. (2017). Introduction to the human gut microbiota. *Biochemical Journal* 474, 1823–1836. https://doi.org/10.1042/BCJ20160510

Tian, W., Wang, C., Gao, Q., Li, L., & Luan, S. (2020). Calcium spikes, waves and oscillations in plant development and biotic interactions. *Nature Plants* 6, 750–759. https://doi.org/10.1038/s41477-020-0667-6

Timsit, Y. & Bennequin, D. (2019). Nervous-like circuits in the ribosome: facts, hypotheses and perspectives. *International Journal of Molecular Sciences* 20, 2911. https://doi.org/10.3390/ijms20122911

Timsit, Y. & Grégoire, S. P. (2021). Towards the idea of molecular brains. *International Journal of Molecular Sciences* 22, 11868. https://doi.org/10.3390/ijms222111868

Timsit, Y., Sergeant-Perthuis, G., & Bennequin, D. (2021). Evolution of ribosomal protein network architectures. *Scientific Reports* 11, 625. https://doi.org/10.1038/s41598-020-80194-4

Togo, T., Alderton, J. M., Bi, G. Q., & Steinhardt, R. A. (1999). The mechanism of facilitated cell membrane resealing. *Journal of Cell Science* 112, 719–731. https://doi.org/10.1242/jcs.112.5.719

Tomasik, B. (2017). Cognitive abilities of unicellular organisms: a compilation. https://reducing-suffering.org/cognitive-abilities-unicellular-organisms/#Summary

Tononi, G. (2004). An information integration theory of consciousness. *BMC Neuroscience* 5, 1–22. https://doi.org/10.1186/1471-2202-5-42

Tononi, G. (2008). Consciousness as integrated information: a provisional manifesto. *The Biological Bulletin* 215, 216–242. https://doi.org/10.2307/25470707

Torday, J. S. (2016). The cell as the first niche construction. *Biology* 5, 19. https://doi.org/10.3390/biology5020019

Torday, J. S. & Miller, W. B., Jr. (2016a). The unicellular state as a point source in a quantum biological system. *Biology* 5, 25. https://doi.org/10.3390/biology5020025

Torday, J. S. & Miller, W. B. (2016b). Biologic relativity: Who is the observer and what is observed?. *Progress in Biophysics and Molecular Biology* 121(1), 29–34.

Torday, J. S. & Miller, W. B., Jr. (2017). The resolution of ambiguity as the basis for life: a cellular bridge between Western reductionism and Eastern holism. *Progress in Biophysics and Molecular Biology* 131, 288–297. https://doi.org/10.1016/j.pbiomolbio.2017.07.013

Torday, J. S. & Miller, W. B., Jr. (2020). *Cellular-Molecular Mechanisms in Epigenetic Evolutionary Biology*. Cham: Springer Nature.

Torday, J. S. & Rehan, V. K. (2012). *Evolutionary Biology, Cell-Cell Communication and Complex Disease*. Hoboken, NJ: Wiley.

Toyofuku, M., Nomura, N., & Eberl, L. (2019). Types and origins of bacterial membrane vesicles. *Nature Reviews Microbiology* 17, 13–24. https://doi.org/10.1038/s41579-018-0112-2

Tozzi, A. & Peters, J. F. (2017). What is it like to be 'the same'? *Progress in Biophysics and Molecular Biology* 133, 30–35.

Trefil, J., Morowitz, H. J., & Smith, E. (2009). The origin of life. A case is made for the descent of electrons. *American Scientist* 97, 206–213. https://www.jstor.org/stable/27859328

Trewavas, A. J. (2014). *Plant Behaviour and Intelligence*. Oxford: Oxford University Press.

Trewavas, A. J. & Baluška, F. (2011). The ubiquity of consciousness: the ubiquity of consciousness, cognition and intelligence in life. *EMBO Reports* 12, 1221–1225. https://doi.org/10.1038/embor.2011.218

Trinh, M. K., Wayland, M. T., & Prabakaran, S. (2019). Behavioural analysis of single-cell aneural ciliate, Stentor roeseli, using machine learning approaches. *Journal of Royal Society Interface* 16, 20190410. https://doi.org/10.1098/rsif.2019.0410

Tsai, M. Y., Zheng, W., Balamurugan, D., Schafer, N. P., Kim, B. L., Cheung, M. S., et al. (2015). Electrostatics, structure prediction, and the energy landscapes for protein folding and binding. *Protein Science* 25, 255.269. https://doi.org/10.1002/pro.2751

Tseng, A. & Levin, M. (2013). Cracking the bioelectric code: probing endogenous ionic controls of pattern formation. *Communicative & Integrative Biology* 6, e22595. https://doi.org/10.4161/cib.2259

Tsuchiya, H. (2017). Anesthetic agents of plant origin: a review of phytochemicals with anesthetic activity. *Molecules* 22, 1369. https://doi.org/10.3390/molecules22081369

Tung, S. L., Boardman, D. A., Sen, M., Letizia, M., Peng, Q., Cianci, N., et al. (2018). Regulatory T cell-derived extracellular vesicles modify dendritic cell function. *Scientific Reports* 8, 6065. https://doi.org/10.1038/s41598-018-24531-8

Turin, L. & Skoulakis, E. M. (2018). Electron spin resonance (EPR) in *Drosophila* and general anesthesia. *Methods in Enzymology* 603, 115–128. https://doi.org/10.1016/bs.mie.2018.01.020

Turin, L., Skoulakis, E. M., & Horsfield, A. P. (2014). Electron spin changes during general anesthesia in *Drosophila*. *Proceedings of the National Academy of Sciences of USA* 111, E3524–E3533. https://doi.org/10.1073/pnas.1404387111

Tuszyński, J. A. (2019). The bioelectric circuitry of the cell. In: Makarov, S., Noetscher, G. M., & Nummenmaa, A. (Eds.), *Brain and Human Body Modeling*, pp. 195–208. Cham: Springer.

Tuszyński, J. A., Portet, S., Dixon, J. M., Luxford, C., & Cantiello, H. F. (2004). Ionic wave propagation along actin filaments. *Biophysical Journal* 86, 1890–1903. https://doi.org/10.1016/S0006-3495(04)74255-1

Ueki, N., Ide, T., Mochiji, S., Kobayashi, Y., Tokutsu, R., Ohnishi, N., et al. (2016). Eyespot-dependent determination of the phototactic sign in *Chlamydomonas reinhardtii*. *Proceedings of the National Academy of Sciences USA* 113, 5299–5304. https://doi.org/10.1073/pnas.1525538113

Uesono, Y. (2009). Environmental stresses and clinical drugs paralyze a cell. *Communicative & Integrative Biology* 2, 275–278. https://doi.org/10.4161/cib.2.3.8226

Uesono, Y., Araki, T., & Toh-E, A. (2008). Local anesthetics, antipsychotic phenothiazines, and cationic surfactants shut down intracellular reactions through membrane perturbation in yeast. *Bioscience, Biotechnology and Biochemistry* 72, 2884–2894. https://doi.org/10.1271/bbb.80385

Uesono, Y., Toh-E, A., Kikuchi, Y., Araki, T., Hachiya, T., Watanabe, C. K., et al. (2016). Local anesthetics and antipsychotic phenothiazines interact nonspecifically with membranes and inhibit hexose transporters in yeast. *Genetics* 202, 997–1012. https://doi.org/10.1534/genetics.115.183806

Umen, J. G. (2014). Green algae and the origins of multicellularity in the plant kingdom. *Cold Spring Harbor Perspectives in Biology* 6, a016170. https://doi.org/10.1101/cshperspect.a016170

Umen, J. & Herron, M. D. (2021). Green algal models for multicellularity. *Annual Review of Genetics* 55, 603–632. https://doi.org/10.1146/annurev-genet-032321-091533

Underhill, D. M., Gordon, S., Imhof, B. A., Núñez, G., & Bousso, P. (2016). Élie Metchnikoff (1845–1916): celebrating 100 years of cellular immunology and beyond. *Nature Reviews Immunology* 16, 651–656. https://doi.org/10.1038/nri.2016.89

Urban, B. W. (2008). The site of anesthetic action. *Handbook of Experimental Pharmacology* 182, 3–29. https://doi.org/10.1007/978-3-540-74806-9_1

Urban, B. W. & Bleckwenn, M. (2002). Concepts and correlations relevant to general anaesthesia. *British Journal of Anaesthesia* 89, 3–16. https://doi.org/10.1093/bja/aef164

Urbiola-Salvador, V., Miroszewska, D., Jabłońska, A., Qureshi, T., & Chen, Z. (2022). Proteomics approaches to characterize the immune responses in cancer. *Biochimica et Biophysica Acta (BBA)-Molecular Cell Research* 1869, 119266. https://doi.org/10.1016/j.bbamcr.2022.119266

Vadakkan, K. I. (2015). A pressure-reversible cellular mechanism of general anesthetics capable of altering a possible mechanism for consciousness. *SpringerPlus* 4, 485. https://doi.org/10.1186/s40064-015-1283-1

Valansi, C., Moi, D., Leikina, E., Matveev, E., Graña, M., Chernomordik, L. V., et al. (2017). *Arabidopsis* HAP2/GCS1 is a gamete fusion protein homologous to somatic and viral fusogens. *Journal of Cell Biology* 216, 571–581. https://doi.org/10.1083/jcb.201610093

Vallejo-Marín, M. (2019). Buzz pollination: studying bee vibrations on flowers. *New Phytologist* 224, 1068–1074. https://doi.org/10.1111/nph.15666

Vallverdú, J., Castro, O., Mayne, R., Talanov, M., Levin, M., Baluška, F., et al. (2017). Slime mould: the fundamental mechanisms of biological cognition. *Biosystems* 165, 57–70. https://doi.org/10.1016/j.biosystems.2017.12.011

van den Berg, J., Boersma, A. J., & Poolman, B. (2017). Microorganisms maintain crowding homeostasis. *Nature Reviews Microbiology* 15, 309–318. https://doi.org/10.1038/nrmicro.2017.17

van Oppen, M. J. H. & Medina, M. (2020). Coral evolutionary responses to microbial symbioses. *Philosophical Transactions of the Royal Society London B Biological Sciences* 375, 20190591. https://doi.org/10.1098/rstb.2019.0591

van Stevendaal, M. H. M. E., Vasiukas, L., Yewdall, N. A., Mason, A. F., & van Hest, J. C. M. (2021). Engineering of biocompatible coacervate-based synthetic cells. *ACS Applied Materials and Interfaces* 13, 7879–7889. https://doi.org/10.1021/acsami.0c19052

Varn, D. P. & Crutchfield, J. P. (2016). What did Erwin mean? The physics of information from the materials genomics of aperiodic crystals and water to molecular information catalysts and life. *Philosophical Transactions of the Royal Society A—Mathematical, Physical and Engineering Sciences* 374, 20150067. https://doi.org/10.1098/rsta.2015.0067

Vastumian, R. D. (2013). The Maxwell demon is on the details. *Physics Today* 69, 49. https://doi.org/10.1063/PT.3.3250

Vaughan, C. V. (2006). Stressed-out endogenous cannabinoids relieve pain. *Trends in Pharmacological Sciences* 27, 69–71. https://doi.org/10.1016/j.tips.2005.11.011

Vaughn, K. C., Renzaglia, K. S. (2006). Structural and immunocytochemical characterization of the *Ginkgo biloba* L. sperm motility apparatus. *Protoplasma* 227, 165–173. https://doi.org/10.1007/s00709-005-0141-3

Veech, R. L., Kashiwaya, Y., Gates, D. N., King, M. T., & Clarke, K. (2002). The energetics of ion distribution: the origin of the resting electric potential of cells. *IUBMB Life* 54, 241–252. https://doi.org/10.1080/15216540215678

Veech, R. L., King, M. T., Pawlosky, R., Bradshaw, P. C., & Curtis, W. (2019). Relationship between inorganic ion distribution, resting membrane potential, and the Δ G' of ATP hydrolysis: a new paradigm. *FASEB Journal* 33, 13126–13130. https://doi.org/10.1096/fj.201901942R

Veits, M., Khait, I., Obolski, U., Zinger, E., Boonman, A., Goldshtein, A., et al. (2019). Flowers respond to pollinator sound within minutes by increasing nectar sugar concentration. *Ecology Letters* 22, 1483–1492. https://doi.org/10.1111/ele.13331

Venn, A. A., Loram, J. E., & Dougalas, A. E. (2008). Photosynthetic symbiosis in animals. *Journal of Experimental Botany* 59, 1069–1080. https://doi.org/10.1093/jxb/erm328

Villarreal, L. P. & Ryan, F. (2018). Viruses in the origin of life and its subsequent diversification. In: Kolb, V. M. (Ed.), *Handbook of Astrobiology*, Chapter 8.3. *Boca Raton, FL*: CRC Press.

Villarreal, L. P. & Witzany, G. (2019). That is life: communicating RNA networks from viruses and cells in continuous interaction. *Annals of the New York Academy of Sciences* 1447, 5–20. https://doi.org/10.1111/nyas.14040

Visick, K. L. & Fuqua, C. (2005). Decoding microbial chatter: cell-cell communication in bacteria. *Journal of Bacteriology* 187, 5507–5519. https://doi.org/10.1128/JB.187.16.5507-5519.2005

Volkmann, D. (2013). Eduard Strasburger. *Biologie in Unserer Zeit* 43, 118–124. https://doi.org/10.1002/biuz.201310505

Volkov, A. G. & Shtessel, Y. B. (2020). Underground electrotonic signal transmission between plants. *Communicative & Integrative Biology* 13, 54–58. https://doi.org/10.1080/19420889.2020.1757207

Vos, M. H., Rappaport, J., Lambry, J. C., Breton, J., & Martin, J. L. (1993). Visualization of coherent nuclear motion in a membrane protein by femtosecond laser spectroscopy. *Nature* 363, 320–325. https://doi.org/10.1038/363320a0

Wächtershäuser, G. (1988). Before enzymes and templates: theory of surface metabolism. *Microbiological Reviews* 52, 452–484. https://doi.org/10.1128/mr.52.4.452-484.1988

Waddington, C. H. (1961). Genetic assimilation. *Advances in Genetics* 10, 257–293. https://doi.org/10.1016/s0065-2660(08)60119-4

Wagner, G. P. & Lynch, V. J. (2010). Evolutionary novelties. *Current Biology* 20, R48–R52. https://doi.org/10.1016/j.cub.2009.11.010

Wagner, M. (2015). Conductive consortia. *Nature* 526, 513–514. https://doi.org/10.1038/526513a

Walker, S. I. & Davies, P. C. (2017). The 'hard problem'of life. In: Walker, S. I., Davies, P. C., & Ellis, G. F. R. (Eds.), *From Matter to Life: Information and Causality*, pp. 19–37. Cambridge: Cambridge University Press.

Walker, S. I., Kim, H., & Davies, P. C. (2016). The informational architecture of the cell. *Philosophical Transactions of the Royal Society A: Mathematical, Physical and Engineering Sciences* 374, 20150057. https://doi.org/10.1098/rsta.2015.0057

Wang, N., Zannoni, N., Ernle, L., Bekö, G., Bekö, G., Wargocki, P., Li, M., et al. (2021). Total OH reactivity of emissions from humans: *in situ* measurement and budget analysis. *Environmental Science & Technology* 55, 149–159. https://doi.org/10.1021/acs.est.0c04206

Wang, S. & Huang, J. (2021). Fungal genes in the innovation and evolution of land plants. *Plant Signaling & Behavior* 16, e1879534. https://doi.org/10.1080/15592324.2021.1879534

Warren, I. A., Naville, M., Chalopin, D., Levin, P., Berger, C. S., Galiana, D., & Volff, J. N. (2015). Evolutionary impact of transposable elements on genomic diversity and lineage-specific innovation in vertebrates. *Chromosome Research* 23, 505–531. https://doi.org/10.1007/s10577-015-9493-5

Warshel, A. & Aqvist, J. (1991). Electrostatic energy and macromolecular function. *Annual Review of Biophysics and Biophysical Chemistry* 20, 267–298. https://doi.org/10.1146/annurev.bb.20.060191.001411

Wegener, G., Krukenberg, V., Riedel, D. Tegetmeyer, H. E., & Boetius, A. (2015). Intercellular wiring enables electron transfer between methanotrophic archaea and bacteria. *Nature* 526, 587–590. https://doi.org/10.1038/nature15733

Wegner, D. M. (2002). *The Illusion of Conscious Will*. Cambridge, MA: MIT Press.

Weinrich, M. & Worcester, D. L. (2013). Xenon and other volatile anesthetics change domain structure in model lipid raft membranes. *Journal of Physical Chemistry B* 117, 16141–16147. https://doi.org/10.1021/jp411261g

Weinrich, M. & Worcester, D. L. (2018). The actions of volatile anesthetics: a new perspective. *Acta Crystallography D Structural Biology* 74, 1169–1177. https://doi.org/10.1107/S2059798318004771

Weir, A. S. A., Chappell, J., & Kacelnik, A. (2002). Shaping of hooks in New Caledonian crows. *Science* 297, 981. https://doi.org/10.1126/science.1073433

Weis, V. M. (2019). Cell biology of coral symbiosis: foundational study can inform solutions to the coral reef crisis. *Integrative & Comparative Biology* 59, 845–855. https://doi.org/10.1093/icb/icz067

Weisenseel, M. H., Niccitelli, R., & Jaffe, L. F. (1975). Large electrical currents traverse growing pollen tubes. *Journal of Cell Biology* 66, 556–567. https://doi.org/10.1083/jcb.66.3.556

White, J. & Yamashita, F. (2022). *Boquila trifoliolata* mimics leaves of an artificial plastic host plant. *Plant Signaling & Behavior* 17, 1977530. https://doi.org/10.1080/15592324.2021.1977530

Whittingstall, K. & Logothetis, N. K. (2009). Frequency-band coupling in surface EEG reflects spiking activity in monkey visual cortex. *Neuron* 64, 281–289. https://doi.org/10.1016/j.neuron.2009.08.016

Wibowo, J. T., Ahmadi, P., Rahmawati, S. I., Bayu, A., Putra, M. Y., & Kijjoa, A. (2021). Marine-derived indole alkaloids and their biological and pharmacological activities. *Marine Drugs* 20, 3. https://doi.org/10.3390/md20010003

Williams, S. D. & Patterson, M. R. (2020). Resistance and robustness of the global coral-symbiont network. *Ecology* 101, e02990. https://doi.org/10.1002/ecy.2990

Wilson, T. H. & Lin, E. C. (1980). Evolution of membrane bioenergetics. *Journal of Supramolecular Structure* 13, 421–446. https://doi.org/10.1002/jss.400130403

Wink, M. (2018). Plant secondary metabolites modulate insect behavior—steps toward addiction? *Frontiers in Physiology* 9, 364. https://doi.org/10.3389/fphys.2018.00364

Witzany, G. (Ed.) (2009). *Natural Genetic Engineering and Natural Genome Editing.* Oxford: Blackwell.

Witzany, G. (2011). The agents of natural genome editing. *Journal of Molecular Cell Biology* 3, 181–189. https://doi.org/10.1093/jmcb/mjr005

Witzany, G. (2012). *Biocommunication of plants.* Springer: Berlin Heidelberg.

Witzany, G. (2014). Pragmatic turn in biology: from biological molecules to genetic content operators. *World Journal of Biological Chemistry* 5, 279. https://doi.org/10.4331/wjbc.v5.i3.279

Witzany, G. (2020). Evolution of genetic information without error replication. In: Burgin, M. & Dodig-Crnkovic, G. (Eds.), *Theoretical Information Studies: Information in the World,* pp. 295–320. Singapore: World Scientific.

Woese, C. (1998). The universal ancestor. *Proceedings of the National Academy of Sciences USA* 95, 6854–6859. https://doi.org/10.1073/pnas.95.12.6854

Woese, C. R. (2002). On the evolution of cells. *Proceedings of the National Academy of Sciences USA* 99, 8742–8747. https://doi.org/10.1073/pnas.132266999

Woese, C. R. (2004). A new biology for a new century. *Microbiology and Molecular Biology Reviews* 68, 173–186. https://doi.org/10.1128/MMBR.68.2.173-186.2004

Wojciech, J., Kamila, M., & Wojciech, B. (2018). Investigation of the population dynamics within a *Pseudomonas aeruginosa* biofilm using a flow based biofilm model system and flow cytometric evaluation of cellular physiology. *Biofouling* 34, 835–850. https://doi.org/10.1080/08927014.2018.1508569

Wood, R. (2003). The ecological evolution of reefs. *Annual Review of Ecology and Systematics* 29, 179–206. https://doi.org/10.1146/annurev.ecolsys.29.1.179

Woods, C. B., Spencer, K. A., Jung, S., Worstman, H. M., Ramirez, J. M., Morgan, P. G., et al. (2023). Mitochondrial function and anesthetic sensitivity in the mouse spinal cord. *Anesthesiology* 134, 901–914. https://doi.org/10.1097/ALN.0000000000003794

Worcester, D. L. & Weinrich, M. (2015). Hydrostatic pressure promotes domain formation in model lipid raft membranes. *Journal of Physical Chemistry Letters* 6, 4417–4421. https://doi.org/10.1021/acs.jpclett.5b02134

Wortzel, I., Dron, S., Kenific, C. M., & Lyden, D. (2019). Exosome-mediated metastasis: communication from a distance. *Developmental Cell* 49, 347–360. https://doi.org/10.1016/j.devcel.2019.04.011

Wright, G. A., Baker, D. D., Palmer, M. J., Stabler, D., Mustard, J. A., Power, E. F., et al. (2013). Caffeine in floral nectar enhances a pollinator's memory of reward. *Science* 339, 1202–1204. https://doi.org/10.1126/science.1228806

Wu, L.-G. & Chan, C. Y. (2022). Multiple roles of actin in exo- and endocytosis. *Frontiers in Synaptic Neuroscience* 14, 841704. https://doi.org/10.3389/fnsyn.2022.841704

Wu, Q., Su, N., Huang, X., Cui, J., Shabala, L., Zhou, M., et al. (2021). Hypoxia-induced increase in GABA content is essential for restoration of membrane potential and preventing ROS-induced disturbance to ion homeostasis. *Plant Communications* 2, 100188. https://doi.org/10.1016/j.xplc.2021.100188

Xie, X., Ding, D., Bai, D., Zhu, Y., Sun, W., Sun, Y., et al. (2022). Melatonin biosynthesis pathways in nature and its production in engineered microorganisms. *Synthetic and Systems Biotechnology* 7, 544–553. https://doi.org/10.1016/j.synbio.2021.12.011

Xu, C., Martin, N., Li, M., & Mann, S. (2022). Living material assembly of bacteriogenic protocells. *Nature* 609, 1029–1037. https://doi.org/10.1038/s41586-022-05223-w

Yamashita, F. & Baluška, F. (2023). Algal ocelloids and plant ocelli. *Plants* 12, 61. https://doi.org/10.3390/plants12010061

Yan, V. T., Narayanan, A., Wiegand, T., Jülicher, F., & Grill, S. W. (2022). A condensate dynamic instability orchestrates actomyosin cortex activation. *Nature* 609, 597–604. https://doi.org/10.1038/s41586-022-05084-3

Yang, C. Y., Bialecka-Fornal, M., Weatherwax, C., Larkin, J. W., Prindle, A., Liu, J., et al. (2020). Encoding membrane-potential-based memory within a microbial community. *Cell Systems* 10, 417–423. https://doi.org/10.1016/j.cels.2020.04.002

Yang, J. & Tavazoie, S. (2020). Regulatory and evolutionary adaptation of yeast to acute lethal ethanol stress. *PLoS One* 15, e0239528. https://doi.org/10.1371/journal.pone.0239528

Yang, Q., Miao, Y., Campanello, L. J., Hourwitz, M. J., Abubaker-Sharif, B., Bull, A. L., et al. (2022). Cortical waves mediate the cellular response to electric fields. *eLife* 11, e73198. https://doi.org/10.7554/eLife.73198

Yang, X. F., Li, L. L., Xu, Y., & Kong, C. H. (2018). Kin recognition in rice (*Oryza sativa*) lines. *New Phytologist* 220, 567–578. https://doi.org/10.1111/nph.15296

Ye, C. -Y., Tang, W., Wu, D., Jia, L., Qiu, J., Chen, M., et al. (2019). Genomic evidence of human selection on Vavilovian mimicry. *Nature Ecology & Evolution* 3, 1474–1482. https://doi.org/10.1038/s41559-019-0976-1

Yeste, M., Jones, C., Amdani, S. N., & Coward, K. (2017). Oocyte activation and fertilisation: crucial contributors from the sperm and oocyte. In: Arur, S. (Ed.), *Signaling-Mediated Control of Cell Division: From Oogenesis to Oocyte-to-Embryo Development*. Repp. 213–239. https://doi.org/10.1007/978-3-319-44820-6_8

Yewdall, N. A. (2022). Life brought to artificial cells. *Nature* 609, 900–901. https://doi.org/10.1038/d41586-022-02231-8

Yewdall, N. M., André, A. A. M., Lu, T., & Spruijt, E. (2021). Coacervates as models of membraneless organelles. *Current Opinion in Colloid & Interface Science* 52, 101416. https://doi.org/10.1016/j.cocis.2020.101416

Yin, Z., Tian, L., Patil, A. J., Li, M., & Mann, S. (2022). Spontaneous membranization in a silk-based coacervate protocell model. *Angewandte Chemie International Edition* 61, e202202302. https://doi.org/10.1002/anie.202202302

Yip, C. Y., Horbour, M. E., Jayawardena, K., Fearnley, I. M., & Sazanov, L. A. (2011). Evolution of respiratory complex I: 'supernumerary' subunits are present in the alpha-proteobacterial enzyme. *Journal of Biological Chemistry* 286, 5023–5033. https://doi.org/10.1074/jbc.M110.194993

Yokawa, K., Kagenishi, T., & Baluška, F. (2019). Anesthetics, anesthesia, and plants. *Trends in Plant Science* 24, 12–14. https://doi.org/10.1016/j.tplants.2018.10.006

Yokawa, K., Kagenishi, T., Pavlovic, A., Gall, S., Weiland, M., Mancuso, S., & Baluška, F. (2018). Anaesthetics stop diverse plant organ movements, affect endocytic vesicle recycling and ROS homeostasis, and block action potentials in Venus flytraps. *Annals of Botany* 122, 747–756. https://doi.org/10.1093/aob/mcx155

Yoon, J. R., Whipple, R. A., Balzer, E. M. Cho, E. H., Matrone, M. A., Peckham, M., et al. (2011). Local anesthetics inhibit kinesin motility and microtentacle protrusions in human epithelial and breast tumor cells. *Breast Cancer Research and Treatment* 129, 691–701. https://doi.org/10.1007/s10549-010-1239-7

Yoshida, M., Muneyuki, E., & Hisabori, T. (2001). ATP synthase—a marvellous rotary engine of the cell. *Nature Reviews Molecular Cell Biology* 2, 669–677. https://doi.org/10.1038/35089509

Yoshida, S., Cui, S., Ichihashi, Y., & Shirasu, K. (2016). The haustorium, a specialized invasive organ in parasitic plants. *Annual Review of Plant Biology* 67, 643–667. https://doi.org/10.1146/annurev-arplant-043015-111702

Yu, H., Schut, G. J., Haja, D. K., & Adams, M. W. W., & Li, H. (2021). Evolution of complex I—like respiratory complexes. *Journal of Biological Chemistry* 296, 100740. https://doi.org/10.1016/j.jbc.2021.100740

Yuryev, M., Pellegrino, C., Jokinen, V., Andriichuk, L., Khirug, S., Khiroug, L., et al. (2016). *In vivo* calcium imaging of evoked calcium waves in the embryonic cortex. *Frontiers in Cellular Neurscience* 9, 500. https://doi.org/10.3389/fncel.2015.00500

Yutin, N., Wolf, M. Y., Wolf, Y. I., & Koonin, E. V. (2009). The origins of phagocytosis and eukaryogenesis. *Biology Direct* 4, 9. https://doi.org/10.1186/1745-6150-4-9

Zaferani, M., Suarez, S. S., & Abbaspourrad, A. (2021). Mammalian sperm hyperactivation regulates navigation via physical boundaries and promotes pseudo-chemotaxis. *Proceedings of National Acadademy of Science USA* 118, e2107500118. https://doi.org/10.1073/pnas.2107500118

Zaremba-Niedzwiedzka, K., Caceres, E. F., Saw, J. H., Bäckström, D., Juzokaite, L., Vancaester, E., et al. (2017). Asgard archaea illuminate the origin of eukaryotic cellular complexity. *Nature* 541, 353–358. https://doi.org/10.1038/nature21031

Žárský, J., Žárský, V., Hanáček, M., & Žárský, V. (2022). Cryogenian glacial habitats as a plant terrestrialisation cradle—the origin of the anydrophytes and zygnematophyceae split. *Frontiers in Plant Science* 12, 735020. https://doi.org/10.3389/fpls.2021.735020

Zamai, L. (2020). Unveiling human non-random genome editing mechanisms activated in response to chronic environmental changes: I. Where might these mechanisms come from and what might they have led to? *Cells* 9, 2362. https://doi.org/10.3390/cells9112362

Zannoni, N., Lakey, P. S. J., Won, Y., Shiraiwa, M., Rim, D., Weschler, C. J., et al. (2022). The human oxidation field. *Science* 377, 1071–1077. https://doi.org/10.1126/science.abn0340

Zerfaß, C., Asally, M., & Soyer, O. S. (2019). Interrogating metabolism as an electron flow system. *Current Opinion in Systems Biology* 13, 59–67. https://doi.org/10.1016/j.coisb.2018.10.001

Zhao, C. & Liu, Y. (2017). Direct interspecies electron transfer between archaea and bacteria. In: Witzany, G. (Ed.), *Biocommunication of Archaea*, pp. 27–40. Ham: Springer International Publishing. https://doi.org/10.1007/978-3-319-65536-9_3

Zhao, D., Yu, Y., Shen, Y., Liu, Q., Zhao, Z., Sharma, R., et al. (2019). Melatonin synthesis and function: evolutionary history in animals and plants. *Frontiers of Endocrinology* 10, 249. https://doi.org/10.3389/fendo.2019.00249

Zhou, Y., Chen, F., Xie, X., Nie, H., Lian, S., Zhong, C., et al. (2021). Tumor-derived exosome promotes metastasis via altering its phenotype and inclusions. *Journal of Cancer* 12, 4240–4246. https://doi.org/10.7150/jca.48043

Zitranski, N., Borth, H., Ackermann, F., Meyer, D., Vieweg, L., Breit, A., et al. (2010). The acrosomal synapse. Subcellular organization by lipid rafts and scaffolding proteins exhibits high similarities in neurons and mammalian spermatozoa. *Communicative & Integrative Biology* 3, 513–521. https://doi.org/10.4161/cib.3.6.13137

Index

For the benefit of digital users, indexed terms that span two pages (e.g., 52–53) may, on occasion, appear on only one of those pages.

The manufacturer's authorised representative in the EU for product safety is Oxford University Press España S.A. of el Parque Empresarial San Fernando de Henares, Avenida de Castilla, 2 – 28830 Madrid (www.oup.es/en or product.safety@oup.com). OUP España S.A. also acts as importer into Spain of products made by the manufacturer.

www.ingramcontent.com/pod-product-compliance
Lightning Source LLC
LaVergne TN
LVHW011753200926
848484LV00003B/694
9780198873211